AF683259

ASTM MANUAL ON PRESENTATION OF DATA AND CONTROL CHART ANALYSIS

Prepared by
Committee E-11 on
Statistical Methods
AMERICAN SOCIETY FOR
TESTING AND MATERIALS

ASTM SPECIAL TECHNICAL PUBLICATION 15D

List Price $9.75
04-015040-34

AMERICAN SOCIETY FOR TESTING AND MATERIALS
1916 Race Street, Philadelphia, Pa. 19103

© by American Society for Testing and Materials 1976
Library of Congress Catalog Card Number: 76-9464

NOTE

The Society is not responsible, as a body,
for the statements and opinions
advanced in this publication.

Printed in Lancaster, Pa.
Oct. 1976

Foreword

ASTM Manual on Presentation of Data and Control Chart Analysis is the fourth revision of the ASTM Manual on Presentation of Data first published in 1933. This fourth revision was prepared by ASTM Committee E-11 on Statistical Methods.

Related ASTM Publications

A Manual on Methods for Retrieving and Correlating Technical Data, STP 468 (1969), $3.00, 04-468000-41

A Note of Appreciation to Reviewers

This publication is made possible by the authors and, also, the unheralded efforts of the reviewers. This body of technical experts whose dedication, sacrifice of time and effort, and collective wisdom in reviewing the papers must be acknowledged. The quality level of ASTM publications is a direct function of their respected opinions. On behalf of ASTM we acknowledge their contribution with appreciation.

ASTM Committee on Publications

Editorial Staff

Jane B. Wheeler, *Managing Editor*
Helen M. Hoersch, *Associate Editor*
Ellen J. McGlinchey, *Assistant Editor*

Contents

SUPPLEMENTS

PART 2
PRESENTING PLUS OR MINUS LIMITS OF UNCERTAINTY OF AN OBSERVED AVERAGE

SUPPLEMENTS

PART 3
CONTROL CHART METHOD OF ANALYSIS AND PRESENTATION OF DATA

GENERAL PRINCIPLES

CONTROL—NO STANDARD GIVEN

CONTROL WITH RESPECT TO A GIVEN STANDARD

CONTROL CHARTS FOR INDIVIDUALS

EXAMPLES

SUPPLEMENTS

APPENDIXES

[a] Available as a separate reprint from ASTM Headquarters.

Preface

This *Manual on the Presentation of Data and Control Chart Analysis, ASTM STP 15D*, was prepared by ASTM Committee E-11 on Statistical Methods to make available to the ASTM membership, and others, information regarding statistical and quality control methods and to make recommendations for their application in engineering work of the Society. The quality control methods considered herein are those methods that have been developed on a statistical basis to control the quality of product through the proper relation of specification, production, and inspection as parts of a continuing process.

The purposes for which the Society was founded—the promotion of knowledge of the materials of engineering, and the standardization of specifications and the methods of testing—involve at every turn the collection, analysis, interpretation, and presentation of quantitative data. Such data form an important part of the source material used in arriving at new knowledge and in selecting standards of quality and methods of testing that are adequate, satisfactory, and economic, from the standpoints of the producer and the consumer.

Broadly, the three general objects of gathering engineering data are to discover: (1) physical constants and frequency distributions, (2) relationships—both functional and statistical—between two or more variables, and (3) causes of observed phenomena. Under these general headings, the following more specific objectives in the work of ASTM may be cited: (*a*) to discover the distributions of quality characteristics of materials which serve as a basis for setting economic standards of quality, for comparing the relative merits of two or more materials for a particular use, for controlling quality at desired levels, for predicting what variations in quality may be expected in subsequently produced material; to discover the distributions of the errors of measurement for particular test methods, which serve as a basis for comparing the relative merits of two or more methods of testing, for specifying the precision and accuracy of standard tests, for setting up economical testing and sampling procedures; (*b*) to discover the relationship between two or more properties of a material, such as density and tensile strength; and (*c*) to discover physical causes of the behavior of materials under particular service conditions; to discover the causes of nonconformance with specified standards in order to make possible the elimination of assignable causes and the attainment of economic control of quality.

Problems falling in these categories can be treated advantageously by the application of statistical methods and quality control methods. This Manual limits itself to several of the items mentioned under (*a*). Part 1 discusses frequency distributions, simple statistical measures, and the presentation, in concise form, of the essential information contained in a single set of *n* observations. Part 2 discusses the problem of expressing ± limits of uncertainty for various statistical measures, together with some working rules for rounding-off observed results to an appropriate number of significant figures. Part 3 discusses the control chart method for the analysis of observational data obtained from a series of samples, and for detecting lack of statistical control of quality.

The present Manual is the fourth revision of earlier work on the subject. The original *ASTM Manual on Presentation of Data, STP 15*, issued in 1933 was prepared by a special committee of former Subcommittee IX on Interpretation and Presentation of Data of ASTM Committee E-1 on Methods of Testing. In 1935, Supplement A on Presenting ± Limits of Uncertainty of an Observed Average and Supplement B on "Control Chart" Method of Analysis and Presentation of Data were issued. These were combined with the original manual and the whole, with minor modifications, was issued as a single volume in 1937. The personnel of the Manual Committee that undertook this early work were: H. F. Dodge, W. C. Chancellor, J. T. McKenzie, R. F. Passano, H. G. Romig, R. T. Webster, and A. E. R. Westman. They were aided in their work by the ready cooperation of the Joint Committee on the Development of Applications of Statistics in Engineering and Manufacturing (sponsored by ASTM and the American Society of Mechanical Engineers (ASME)) and especially of the chairman of the Joint Committee, W. A. Shewhart. The nomenclature and symbolism used in this early work were adopted in 1941 and 1942 in the American War Standards on Quality Control (Z1.1, Z1.2, and Z1.3) of the American Standards Association, and its Supplement B was reproduced as an appendix with one of these standards.

In 1946, ASTM Committee E-11 on Statistical Methods was established under the chairmanship of H. F. Dodge, and the manual became its responsibility. A major revision was issued in 1951 as *ASTM Manual on Quality Control of Materials, STP 15C*. The Task Group that undertook the revision of Part 1 consisted of R. F. Passano, Chairman, H. F. Dodge, A. C. Holman, and J. T. McKenzie. The same task group also revised Part 2 (the old Supplement A) and the task group for revision of Part 3 (the old Supplement B) consisted of A. E. R. Westman, Chairman, H. F. Dodge, A. I. Peterson, H. G. Romig, and L. E. Simon. In this 1951 revision, the term "confidence limits" was introduced and constants for computing 0.95 confidence limits were added to the constants for 0.90 and 0.99 confidence limits presented in prior printings. Separate treatment was given to control charts for "number of defectives," "number of defects," and "number of defects per unit" and material on control charts for individuals was added.

In the 1956 printing of *ASTM STP 15C*, the ASTM Recommended Practice for Choice of Sample Size to Estimate the Average Quality of a Lot or Process (E 122) was included in the Appendix. This recommended practice had been prepared by a task group of ASTM Committee E-11 consisting of A. G. Scroggie, Chairman, A. C. Bicking, W. E. Deming, H. F. Dodge, and S. B. Littauer. In the 1960 printing, a number of minor modifications were made by an ad hoc committee consisting of Harold Dodge, Chairman, Simon Collier, R. H. Ede, R. J. Hader, and E. G. Olds. At this time, mention was made of $s^2 = \Sigma(X - \overline{X})^2/(n - 1)$ as an unbiased estimate of the universe variance and it was indicated that s, though not unbiased, was regarded as an estimate of the universe standard deviation. The root-mean-square deviation $\sigma = \sqrt{\Sigma(X - \overline{X})^2/n}$ was still retained as the definition of the sample standard deviation.

The present manual is a revision and replacement of *ASTM STP 15C*. Although over the years *ASTM STP 15C* has been widely received and has become a classic in the area of quality control, there has been a growing feeling that its usefulness might be enhanced if certain definitions and notations were brought into accord with modern practice and if the discussion were rounded out in spots by the addition of new material. It seemed especially desirable to bring the manual into line with the American Society for Quality Control Standard A1-1971 (ANSI Z1.5-1971).

Definitions, Symbols, Formulas, and Tables for Control Charts

The principal change in *ASTM STP 15C* introduced in the present Manual is the redefinition of the sample standard deviation to be $s = \sqrt{\Sigma(X_i - \overline{X})^2/(n - 1)}$. This change has required numerous changes throughout the Manual in mathematical equations and formulas, tables, and numerical illustrations. It has also led to a sharpening of distinctions between sample values, universe values, and standard values which were not formerly deemed necessary. Among the changes caused by the redefinition of the sample standard deviation has been the introduction of the c_4 factor in place of the c_2 factor, the A_3 factor in place of the A_1 factor, the B_5 and B_6 factors in place of the B_1 and B_2 factors, and the E_3 factor in place of the E_1 factor. These changes are reflected in Tables 2, 3, 27, and 29 and Fig. 1 of Part 3. The mathematical development in Supplement A to Part 3 has also had to be modified.

It should be carefully noted that the redefinition of the sample standard deviation as $s = \sqrt{\Sigma(X - \overline{X})^2/(n - 1)}$ and the resulting need to introduce factors c_4, A_3, B_5, B_6, and E_3 in place of factors c_2, A_1, B_1, B_2, and E_1 has no practical effect on the control chart analysis in which these quantities are used, nor does the redefinition affect the confidence limits that are set up. The change is simply in the statistical functions that are used in computing the results.

The changes in notation in the present Manual unrelated to the redefinition of the sample standard deviation have been the substitution of d for x, of k for t in the discussion of Tchebychev's inequality, of k for t in the discussion of the Normal distribution in Section 32 of Part 1, and of g_1 for k, the sample coefficient of skewness. The symbol σ has been retained for the sample root-mean-square deviation but is not now defined as the sample standard deviation. It is related to s, of course, by the formula $\sigma = s\sqrt{(n-1)/n}$. In Part 3 on control chart analysis, standard values have been distinguished from actual universe values by using a zero subscript. A summary of all changes in symbols is given on p. xix.

The new material added in the present Manual is as follows. The sample measure of kurtosis, g_2, has been introduced. This has led to revision of Table 7 and Section 33. In addition, note is now made of the special aspects of the presentation of data to be considered when the distribution of X follows a negative exponential distribution or a Weibull distribution, both important in the area of reliability and life testing. In Part 2, a brief discussion of the determination of confidence limits for a universe standard deviation and a universe proportion has also been included.

The Task Group responsible for this fourth revision of the Manual consisted of A. J. Duncan, Chairman, R. A. Freund, F. E. Grubbs, and D. C. McCune. All members of ASTM Committee E-11 had an opportunity to review the revised material and the comments received were a significant factor in the improvement of the text.

Summary of Changes in Definitions and Notations

ASTM STP 15D	*ASTM STP 15C*
s (= sample standard deviation, s^2 = sample variance = unbiased estimate of σ'^2)	s (s^2 = unbiased estimate of σ'^2)
σ (= sample root-mean-square deviation)	σ (= sample standard deviation = root-mean-square deviation)
c_4 (s/c_4 or $\bar{s}/c_4$ = unbiased estimate of σ')	c_2 (σ/c_2 or $\bar{\sigma}/c_2$ = unbiased estimate of σ')
A_3 (control limits for $\bar{X} = \bar{\bar{X}} \pm A_3\bar{s}$)	A_1 (control limits for $\bar{X} = \bar{\bar{X}} \pm A_1\bar{\sigma}$)
B_5, B_6 (control limits for s, given standard value σ_0', = $B_5\sigma_0'$ and $B_6\sigma_0'$)	B_1, B_2 (control limits for σ, given standard value σ_0', = $B_1\sigma'$ and $B_2\sigma'$)
E_3 (control limits for individuals $= \bar{\bar{X}} \pm E_3\bar{s}$)	E_1 (control limits for individuals $= \bar{\bar{X}} \pm E_1\sigma$)
$\bar{X}'$, σ', p', μ', c' (= universe values)	$\bar{X}'$, σ', p', μ', c' (= universe values or standard values as text implies)
X_0', σ_0', p_0', μ_0', c_0' (= standard values)	
d (= $X - A$)	x (= $X - A$)
k ($k\sigma'$ = deviation in Tchebychev's inequality or deviation of a normal variable)	t ($t\sigma$ = deviation in Tchebychev's inequality)
	t ($t\sigma'$ = deviation of a normal variable)
t = student's t	
g_1 = sample coefficient of skewness	(sometimes k is used as universe coefficient of skewness but generally stands for the sample coefficient)
g_2 = sample coefficient of kurtosis	

PART 1

Presentation of Data

*Part 1 is concerned solely with presenting information about a given sample of data. It contains no discussion of inferences that might be made about the universe from which the sample came.

Presentation of Data

SUMMARY: Bearing in mind that no rules can be laid down to which no exceptions can be found the committee believes that if the recommendations presented are followed, the presentations will contain the essential information for a majority of the uses made of ASTM data.

RECOMMENDATIONS FOR PRESENTATION OF DATA: Given a sample of n observations of a single variable obtained under the same essential conditions:

1. Present as a minimum, the average, the standard deviation, and the number of observations. *Always* state the number of observations.
2. If the number of observations is large and if it is desired to use a theoretical curve as a graduation formula, present also the values of the skewness, g_1, and the kurtosis, g_2. Alternately, present a grouped frequency distribution.
3. If the data were not obtained under controlled conditions and it is desired to give information regarding the extreme observed effects of assignable causes, present the values of the maximum and minimum observations in addition to the average, the standard deviation, and the number of observations.
4. Present as much evidence as possible that the data were obtained under controlled conditions.
5. Present relevant information on precisely (*a*) the field within which the measurements are supposed to hold and (*b*) the conditions under which they were made.

Glossary of Symbols used in Part 1

- d Used in Section 25 to designate deviation in cells from an arbitrary origin
- f *Observed frequency* (number of observations) in a single cell of a frequency distribution
- g_1 *Sample coefficient of skewness*, a measure of skewness, or lopsidedness of a distribution
- g_2 *Sample coefficient of kurtosis*, see p. 20.
- n Number of observed values (observations)
- p *Sample relative frequency* or *proportion*, the ratio of the number of occurrences of a given type to the total possible number of occurrences, the ratio of the number of observations in any stated interval to the total number of observations; *sample fraction defective*, for measured values the ratio of the number of observations lying outside specified limits (or beyond a specified limit) to the total number of observations
- R *Sample range*, the difference between the largest observed value and the smallest observed value
- s *Sample standard deviation*

s^2 *Sample variance*
σ *Root-mean-square deviation*
v *Sample coefficient of variation*, a measure of relative dispersion based on the standard deviation
X Observed values of a measurable characteristic; specific observed values are designated X_1, X_2, X_3, etc; also used to designate a measurable characteristic
$\bar{X}$ *Sample* average or sample mean, the sum of the n observed values in a sample divided by n

If reference is to be made to the universe from which a given sample came, the following symbols should be used.

γ_1 Universe skewness defined as the expected value (see Note) of $(X - \bar{X}')^3$ divided by σ'^3
γ_2 Universe coefficient of kurtosis defined as the amount by which the expected value (see Note) of $(X - \bar{X}')^4$ divided by σ'^4 exceeds or falls short of 3
p' Universe relative frequency
σ' Universe standard deviation
σ'^2 *Universe variance* defined as the expected value (see Note) of the square of a deviation from the universe mean; thus $E(X - \bar{X}')^2 = \sigma'^2$
V *Universe* coefficient of variation
$\bar{X}'$ *Universe average* or *universe mean* defined as the expected value (see Note) of X; thus $E(X) = \bar{X}'$

NOTE—Mathematically the expected value of a function of X, say $h(X)$, is defined as the sum or integral of that function times the probability of X. For example, if the probability of X lying between X and $X + dX$ is $f(X)\,dX$ then the expected value of $(X - \bar{X}')^2$, for example, is $\int (X - \bar{X}')^2 f(X)\,dX$.

INTRODUCTION

1. Purpose

This Part 1 of the Manual discusses the application of statistical methods to the problem of: (*a*) condensing the information contained in a sample[1] of observations, and (*b*) presenting the essential information in a concise form more readily interpretable than the unorganized mass of original data.

Attention will be directed particularly to quantitative information on measurable characteristics of materials and manufactured products. Such characteristics will be termed *quality characteristics*.

[1] In the 1975 printing, the word "sample" was substituted for the word "set" where it appeared more appropriate.

2. Type of Data Considered

Consideration will be given to the treatment of a sample of n observations of a single variable. Figure 1 illustrates two general types: (*a*) the first type is a series of n observations representing single measurements of the same

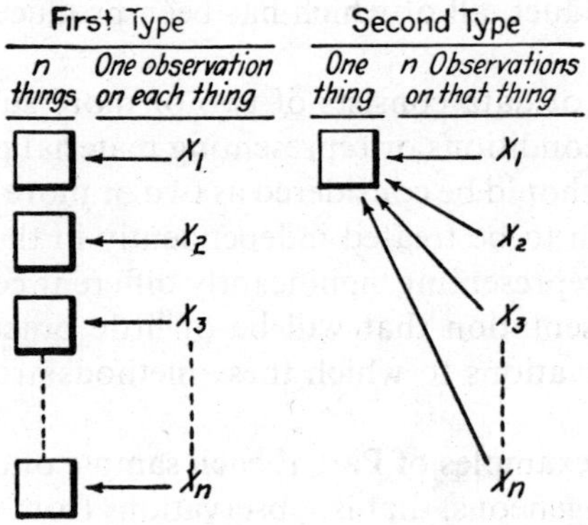

FIG. 1—*Two general types of data.*

quality characteristic of n similar things, and (*b*) the second type is a series of n observations representing n measurements of the same quality characteristic of one thing.

Data of the first type are commonly gathered to furnish information regarding the *distribution* of the quality of the material itself, having in mind possibly some more specific purpose, such as the establishment of a quality standard or the determination of conformance with a specified quality standard, for example, 100 observations of transverse strength on 100 bricks of a given brand.

Data of the second type are commonly gathered to furnish information regarding the errors of measurement for a particular test method, for example, 50 micrometer measurements of the thickness of a test block.

NOTE—The quality of a material in respect to some particular characteristic, such as tensile strength, is a frequency distribution function, not a single-valued constant.

The variability in a group of observed values of such a quality characteristic is made up of two parts: variability of the material itself, and the errors of measurement. In some practical problems, the error of measurement may be large compared with the variability of the material; in others, the converse may be true. In any case, if one is interested in discovering the objective frequency distribution of the quality of the material, consideration must be given to correcting the errors of measurement (Shewhart [*1*][2] pp. 379–385).

3. Homogeneous Data

While the methods here given may be used to condense any set of observations, the results obtained by using them may be of little value from the

[2] The italic numbers in brackets refer to the list of references appended to Part 1.

standpoint of interpretation unless the data are good in the first place and satisfy certain requirements.

To be useful for inductive generalization, any sample of observations that is treated as a single group for presentation purposes should represent a series of measurements, all made under essentially the same test conditions, on a material or product, all of which has been produced under essentially the same conditions.

If a given sample of data consists of two or more subportions collected under different test conditions or representing material produced under different conditions, it should be considered as two or more separate subgroups of observations, each to be treated independently in the analysis. Merging of such subgroups, representing significantly different conditions, may lead to a condensed presentation that will be of little practical value. Briefly, any sample of observations to which these methods are applied should be *homogeneous.*

In the illustrative examples of Part 1, each sample of observations will be assumed to be homogeneous, that is, observations from a common universe of causes. The analysis and presentation by control chart methods of data obtained from several samples or capable of subdivision into subgroups on the basis of relevant engineering information is discussed in Part 3 of this Manual. Such methods enable one to determine whether for practical purposes a given sample of observations may be considered to be homogeneous.

4. Typical Examples of Physical Data

Table 1 gives three typical sets of observations, each representing measurements on a sample of units or specimens selected in a random manner to provide information about the quality of a larger quantity of material—the general output of one brand of brick, a production lot of galvanized iron sheets, and a shipment of hard drawn copper wire. Consideration will be given to ways of arranging and condensing these data into a form better adapted for practical use.

UNGROUPED FREQUENCY DISTRIBUTIONS

5. Ungrouped Frequency Distribution

An arrangement of the observed values in ascending order of magnitude will be referred to in the Manual as the *ungrouped frequency distribution* of the data, to distinguish it from the grouped frequency distribution defined in Section 8. Table 2 presents ungrouped frequency distributions for the three sets of observations given in Table 1.

Figure 2 shows graphically the ungrouped frequency distribution of Table 2 (*a*). In the graph there is a minor grouping in terms of the unit of measurement. For the data of Fig. 2, it is the "rounding-off" unit of 10 psi.

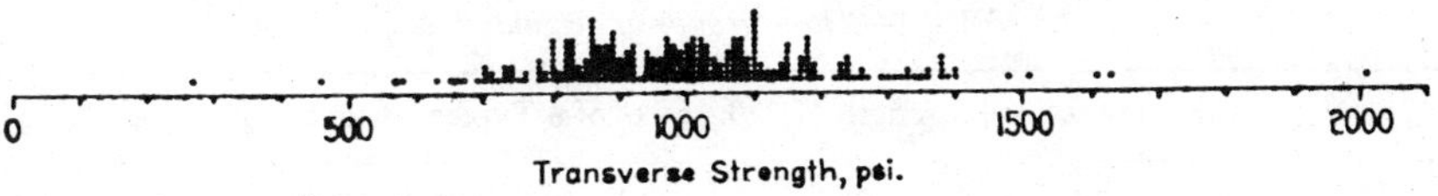

FIG. 2—*Showing graphically the ungrouped frequency distribution of a set of observations. Each dot represents one brick, data of Table 2*(a).

A glance at one of the tabulations of Table 2 gives some information not readily observed in the original data of Table 1—such as the maximum, the minimum, and the median or middlemost value. Such arrangements are sometimes of value in the initial stages of analysis.

6. Remarks

It is rarely desirable to present data in the manner of Table 1 or Table 2. The mind cannot grasp in its entirely the meaning of so many numbers; furthermore, greater compactness is required for most of the practical uses that are made of data.

GROUPED FREQUENCY DISTRIBUTIONS

7. Introduction

The information contained in a set of observations may be condensed merely by grouping. Such grouping involves some loss of information but is often useful in presenting engineering data. In the following sections both tabular and graphical presentation of grouped data will be discussed.

8. Definitions

A *grouped frequency distribution* of a set of observations is an arrangement which shows the frequency of occurrence of the values of the variable in ordered classes.

The interval, along the scale of measurement, of each ordered class is termed a *cell*.

The *frequency* for any cell is the number of observations in that cell.

The *relative frequency* for any cell is the frequency for that cell divided by the total number of observations.

Table 3 illustrates how the three sets of observations given in Table 1 may be organized into grouped frequency distributions. The recommended form of presenting tabular distributions is somewhat more compact, however, as shown in Table 4. Graphical presentation is used in Fig. 3 and discussed in detail in Section 14.

TABLE 1—*Three groups of original data.*

(*a*) Transverse Strength of 270 Bricks of a Typical Brand, psi[a]

860	1320	820	1040	1000	1010	1190	1180	1080	1100	1130
920	1100	1250	1480	1150	740	1080	860	1000	810	1000
1200	830	1100	890	270	1070	830	1380	960	1360	730
850	920	940	1310	1330	1020	1390	830	820	980	1330
920	1070	1630	670	1150	1170	920	1120	1170	1160	1090
1090	700	910	1170	800	960	1020	1090	2010	890	930
830	880	870	1340	840	1180	740	880	790	1100	1260
1040	1080	1040	980	1240	800	860	1010	1130	970	1140
1510	1060	840	940	1110	1240	1290	870	1260	1050	900
740	1230	1020	1060	990	1020	820	1030	860	850	890
1150	860	1100	840	1060	1030	990	1100	1080	1070	970
1000	720	800	1170	970	690	1020	890	700	880	1150
1140	1080	990	570	790	1070	820	580	820	1060	980
1030	960	870	800	1040	820	1180	1350	1180	950	1110
700	860	660	1180	780	1230	950	900	760	1380	900
920	1100	1080	980	760	830	1220	1100	1090	1380	1270
860	990	890	940	910	1100	1020	1380	1010	1030	950
950	880	970	1000	990	830	850	630	710	900	890
1020	750	1070	920	870	1010	1230	780	1000	1150	1360
1300	970	800	650	1180	860	1150	1400	880	730	910
890	1030	1060	1610	1190	1400	850	1010	1010	1240	
1080	970	960	1180	1050	920	1110	780	780	1190	
910	1100	870	980	730	800	800	1140	940	980	
870	970	910	830	1030	1050	710	890	1010	1120	
810	1070	1100	460	860	1070	880	1240	940	860	

(*b*) Weight of Coating of 100 Sheets of Galvanized Iron Sheets, oz/ft^2 [b]

1.467	1.603	1.577	1.563	1.437
1.623	1.603	1.577	1.393	1.350
1.520	1.383	1.323	1.647	1.530
1.767	1.730	1.620	1.620	1.383
1.550	1.700	1.473	1.530	1.457
1.533	1.600	1.420	1.470	1.443
1.377	1.603	1.450	1.337	1.473
1.373	1.477	1.337	1.580	1.433
1.637	1.513	1.440	1.493	1.637
1.460	1.533	1.557	1.563	1.500
1.627	1.593	1.480	1.543	1.607
1.537	1.503	1.477	1.567	1.423
1.533	1.600	1.550	1.670	1.573
1.337	1.543	1.637	1.473	1.753
1.603	1.567	1.570	1.633	1.467
1.373	1.490	1.617	1.763	1.563
1.457	1.550	1.477	1.573	1.503
1.660	1.577	1.750	1.537	1.550
1.323	1.483	1.497	1.420	1.647
1.647	1.600	1.717	1.513	1.690

(*c*) Breaking, Strength of Ten Specimens of 0.104-in. Hard-Drawn Copper Wire, lb[c]

578
572
570
568
572
570
570
572
576
584

[a]Measured to the nearest 10 psi. Test method used was ASTM Standard Methods of Testing Brick (C 67-31). The latest issue of this test method is ASTM Sampling and Testing Brick and Structural Clay Tile (C 67-73). Data from *ASTM Manual for Interpretation of Refractory Test Data*, 1935, p. 83.

[b]Measured to the nearest 0.01 oz/ft^2 of sheet, averaged for three spots. Test method used was ASTM Triple Spot Test of Standard Specifications for Zinc-Coated (Galvanized) Iron or Steel Sheets (A 93-27). The latest issue of this test method is ASTM Specification for Steel Sheet, Zinc-Coated (Galvanized) by the Hot-Dip Process, General Requirements (A 525-73). Data from laboratory tests.

[c]Measured to the nearest 2 lb. Test method used was ASTM Standard Specifications for Hard-Drawn Copper Wire (B 1-27). The latest issue of this test method is B 1-70. Data from inspection report.

TABLE 2—*Ungrouped frequency distributions in tabular form.*

(*a*) Transverse Strength, psi (data of Table 1 (*a*))

270	780	830	870	920	970	1020	1070	1100	1180	1310
460	780	830	880	920	980	1020	1070	1100	1180	1320
570	780	830	880	920	980	1020	1070	1100	1180	1330
580	790	840	880	920	980	1020	1070	1100	1180	1330
630	790	840	880	920	980	1020	1070	1110	1180	1340
650	800	840	880	930	980	1020	1070	1110	1180	1350
660	800	850	880	940	980	1020	1070	1110	1180	1360
670	800	850	890	940	990	1030	1080	1120	1190	1360
690	800	850	890	940	990	1030	1080	1120	1190	1380
700	800	850	890	940	990	1030	1080	1130	1190	1380
700	800	860	890	940	990	1030	1080	1130	1200	1380
700	800	860	890	950	990	1030	1080	1140	1220	1380
710	810	860	890	950	1000	1030	1080	1140	1230	1390
710	810	860	890	950	1000	1040	1080	1140	1230	1400
720	820	860	890	950	1000	1040	1090	1150	1230	1400
730	820	860	900	960	1000	1040	1090	1150	1240	1480
730	820	860	900	960	1000	1040	1090	1150	1240	1510
730	820	860	900	960	1000	1050	1090	1150	1240	1610
740	820	860	900	960	1010	1050	1100	1150	1240	1630
740	820	860	910	970	1010	1050	1100	1150	1250	2010
740	820	870	910	970	1010	1060	1100	1160	1260	
750	830	870	910	970	1010	1060	1100	1170	1260	
760	830	870	910	970	1010	1060	1100	1170	1270	
760	830	870	910	970	1010	1060	1100	1170	1290	
780	830	870	920	970	1010	1060	1100	1170	1300	

TABLE 2—*Continued.*

(*b*) Weight of Coating, oz/ft² (data of Table 1 (*b*))					(*c*) Breaking Strength, lb. (data of Table 1 (*c*))
1.323	1.457	1.513	1.567	1.620	568
1.323	1.457	1.513	1.567	1.623	570
1.337	1.460	1.520	1.570	1.627	570
1.337	1.467	1.530	1.573	1.633	570
1.337	1.467	1.530	1.573	1.637	572
1.350	1.470	1.533	1.577	1.637	572
1.373	1.473	1.533	1.577	1.637	572
1.373	1.473	1.533	1.577	1.647	576
1.377	1.473	1.537	1.580	1.647	578
1.383	1.477	1.537	1.593	1.647	584
1.383	1.477	1.543	1.600	1.660	
1.393	1.477	1.543	1.600	1.670	
1.420	1.480	1.550	1.600	1.690	
1.420	1.483	1.550	1.603	1.700	
1.423	1.490	1.550	1.603	1.717	
1.433	1.493	1.550	1.603	1.730	
1.437	1.497	1.557	1.603	1.750	
1.440	1.500	1.563	1.607	1.753	
1.443	1.503	1.563	1.617	1.763	
1.450	1.503	1.563	1.620	1.767	

TABLE 3—*Three examples of grouped frequency distribution, showing cell midpoints and cell boundaries.*

	Cell Midpoint	Cell Boundaries	Observed Frequency
(*a*) Transverse strength, psi (data of Table 1 (*a*))	...	225	...
	300	...	1
	...	375	...
	450	...	1
	...	525	...
	600	...	6
	...	675	...
	750	...	38
	...	825	...
	900	...	80
	...	975	...
	1050	...	83
	...	1125	...
	1200	...	39
	...	1275	...
	1350	...	17
	...	1425	...
	1500	...	2

TABLE 3—*Continued.*

	Cell Midpoint	Cell Boundaries	Observed Frequency
	...	1575	...
	1650	...	2
	...	1725	...
	1800	...	0
	...	1875	...
	1950	...	1
	...	2025	...
		total	270
(*b*) Weight of coating,	...	1.275	...
oz/ft^2 (data of Table 1 (*b*))	1.300	...	2
	...	1.325	...
	1.350	...	6
	...	1.375	...
	1.400	...	7
	...	1.425	...
	1.450	...	14
	...	1.475	...
	1.500	...	14
	...	1.525	...
	1.550	...	22
	...	1.575	...
	1.600	...	17
	...	1.625	...
	1.650	...	10
	...	1.675	...
	1.700	...	3
	...	1.725	...
	1.750	...	5
	...	1.775	...
		total	100
(*c*) Breaking strength, lb	...	567	...
(data of Table 1 (*c*))	568	...	1
	...	569	...
	570	...	3
	...	571	...
	572	...	3
	...	573	...
	574	...	0
	...	575	...
	576	...	1
	...	577	...
	578	...	1
	...	579	...
	580	...	0
	...	581	...
	582	...	0
	...	583	...
	584	...	1
	...	585	...
		total	10

TABLE 4—*Four methods of presenting a tabular frequency distribution (data of Table 1 (a)).*

(*a*) Frequency

Transverse Strength, psi	Number of Bricks Having Strength Within Given Limits
225 to 375	1
375 to 525	1
525 to 675	6
675 to 825	38
825 to 975	80
975 to 1125	83
1125 to 1275	39
1275 to 1425	17
1425 to 1575	2
1575 to 1725	2
1725 to 1875	0
1875 to 2025	1
Total	270

(*b*) Relative Frequency (expressed in percentages)

Transverse Strength, psi	Percentage of Bricks Having Strength Within Given Limits
225 to 375	0.4
375 to 525	0.4
525 to 675	2.2
675 to 825	14.1
825 to 975	29.6
975 to 1125	30.7
1125 to 1275	14.5
1275 to 1425	6.3
1425 to 1575	0.7
1575 to 1725	0.7
1725 to 1875	0.0
1875 to 2025	0.4
Total	100.0

number of observations = 270

(*c*) Cumulative Frequency

Transverse Strength, psi	Number of Bricks Having Strength less than Given Values
375	1
525	2
675	8
825	46
975	126
1125	209
1275	248
1425	265
1575	267
1725	269
1875	269
2025	270

(*d*) Cumulative Relative Frequency (expressed in percentages)

Transverse Strength, psi	Percentage of Bricks Having Strength less than Given Values
375	0.4
525	0.8
675	3.0
825	17.1
975	46.7
1125	77.4
1275	91.9
1425	98.2
1575	98.9
1725	99.6
1875	99.6
2025	100.0

number of observations = 270

NOTE—"Number of observations" should be recorded with tables of relative frequencies.

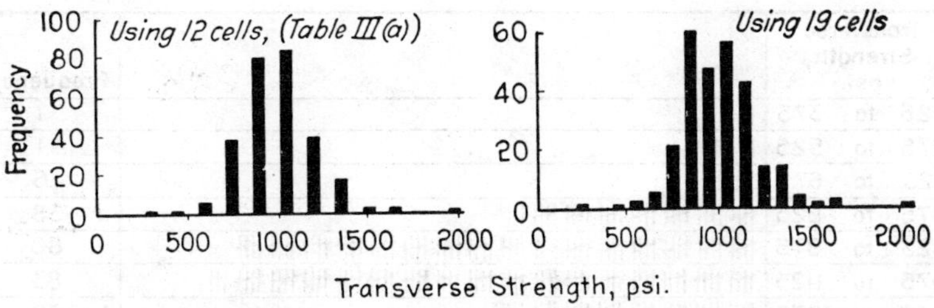

FIG. 3—*Illustrating increased irregularity with larger number of cells.*

9. Choice of Cell Boundaries

It is usually advantageous to make the cell intervals equal.

It is recommended that, in general, the *cell boundaries* be chosen half-way between two possible observations. By choosing cell boundaries in this way, certain difficulties of classification and computation are avoided [*2*]. With this choice, the cell boundary values will usually have one more significant figure (usually a 5) than the values in the original data. For example, in Table 3 (*a*), observations were recorded to the nearest 10 psi, hence the cell boundaries were placed at 225, 375, etc., rather than at 220, 370, etc., or 230, 380, etc. Likewise, in Table 3 (*b*), observations were recorded to the nearest 0.01 oz/ft^2, hence cell boundaries were placed at 1.275, 1.325, etc., rather than at 1.28, 1.33, etc.

10. Number of Cells

The number of cells in a frequency distribution should preferably be between 13 and 20. (For a discussion of this point, see Ref *1*, p. 69.) If the number of observations is, say, less than 250, as few as 10 cells may be of use. When the number of observations is less than 25, a frequency distribution of the data is generally of little value from a presentation standpoint, as for example the 10 observations in Table 3 (*c*). In general, the outline of a frequency distribution when presented graphically is more irregular, the larger the number of cells. This tendency is illustrated in Fig. 3.

11. Methods of Classifying Observations

Figure 4 illustrates a convenient method of classifying observations into cells when the number of observations is not large. For each observation, a mark is entered in the proper cell. These marks are grouped in fives as the tallying proceeds, and the completed tabulation itself, if neatly done, provides a good picture of the frequency distribution.

Transverse Strength, psi.		Frequency
225 to 375	|	1
375 to 525	|	1
525 to 675	𝍸 |	6
675 to 825	𝍸 𝍸 𝍸 𝍸 𝍸 𝍸 𝍸 |||	38
825 to 975	𝍸 𝍸 𝍸 𝍸 𝍸 𝍸 𝍸 𝍸 𝍸 𝍸 𝍸 𝍸 𝍸 𝍸 𝍸 𝍸	80
975 to 1125	𝍸 𝍸 𝍸 𝍸 𝍸 𝍸 𝍸 𝍸 𝍸 𝍸 𝍸 𝍸 𝍸 𝍸 𝍸 𝍸 |||	83
1125 to 1275	𝍸 𝍸 𝍸 𝍸 𝍸 𝍸 𝍸 ||||	39
1275 to 1425	𝍸 𝍸 𝍸 ||	17
1425 to 1575	||	2
1575 to 1725	||	2
1725 to 1875		0
1875 to 2025	|	1
	Total	270

FIG. 4—*Method of classifying observations. Data of Table 1*(a).

If the number of observations is, say, over 250, and accuracy is essential, it may be found advantageous to enter the observed values on cards, one to each observation. These may then be sorted into packs, each pack corresponding to a cell. By this means, the work of classification can be checked by making sure that no card has been wrongly sorted. When a large amount of data is to be analyzed, resort may be made to an electronic computer.

12. Cumulative Frequency Distribution

For some purposes, the number of observations having a value "less than" or "greater than" particular scale values is of more importance than the frequencies for particular cells. A table of such frequencies is termed a *cumulative frequency distribution*. The "less than" cumulative frequency distribution is formed by recording the frequency of the first cell, then the sum of the first and second cell frequencies, then the sum of the first, second, and third cell frequencies, and so on.

13. Tabular Presentation

Methods of presenting tabular frequency distributions are shown in Table 4. To make a frequency tabulation more understandable, relative frequencies may be listed as well as actual frequencies. If only relative frequencies are given, the table cannot be regarded as complete unless the total number of observations is recorded.

Confusion often arises from failure to record cell boundaries correctly. Of the four methods, A to D, illustrated for strength measurements made *to the nearest 10 lb,* only Methods A and B are recommended. Method C gives no clue as to how observed values of 2100, 2200, etc., which fell exactly at cell boundaries were classified. If such values were consistently placed in the next higher cell, the real cell boundaries are those of Method A. Method D is liable to misinterpretation since strengths were measured to the nearest 10 lb only.

Recommended				Not Recommended			
Method A		Method B		Method C		Method D	
Strength, lb	Number of Obser-vations	Strength, lb	Number of Obser-vations	Strength, lb	Number of Obser-vations	Strength, lb	Number of Obser-vations
1995 to 2095	1	2000 to 2090	1	2000 to 2100	1	2000 to 2099	1
2095 to 2195	3	2100 to 2190	3	2100 to 2200	3	2100 to 2199	3
2195 to 2295	17	2200 to 2290	17	2200 to 2300	17	2200 to 2299	17
2295 to 2395	36	2300 to 2390	36	2300 to 2400	36	2300 to 2399	36
2395 to 2495	82	2400 to 2490	82	2400 to 2500	82	2400 to 2499	82

14. Graphical Presentation

Using a convenient horizontal scale for values of the variable and a vertical scale for cell frequencies, frequency distributions may be reproduced graphically in several ways as shown in Fig. 5. The *frequency bar chart* is obtained by erecting a series of bars, centered on the cell midpoints, each bar having a height equal to the cell frequency. An alternate form of frequency bar chart may be constructed by using lines rather than bars. The distribution may also be shown by a series of points or circles representing cell frequencies plotted at cell midpoints. The *frequency polygon* is obtained by joining these points by straight lines. Each end point is joined to the base at the next cell midpoint to close the polygon.

Another form of graphical representation of a frequency distribution is obtained by placing along the graduated horizontal scale a series of vertical columns, each having a width equal to the cell width and a height equal to the cell frequency. Such a graph, shown at the bottom of Fig. 5, is called the *frequency histogram* of the distribution. In the histogram, if cell widths are arbitrarily given the value 1, the area enclosed by the steps represents frequency exactly, and the sides of the columns designate cell boundaries.

The same charts can be used to show relative frequencies by substituting a relative frequency scale, such as that shown at the right in Fig. 5. It is often advantageous to show both a frequency scale and a relative frequency scale. If only a relative frequency scale is given on a chart, the number of observations should be recorded.

Two methods of constructing cumulative frequency polygons are shown in Fig. 6. Points are plotted at cell boundaries. The upper chart gives cumulative frequency and relative cumulative frequency plotted on an arithmetic

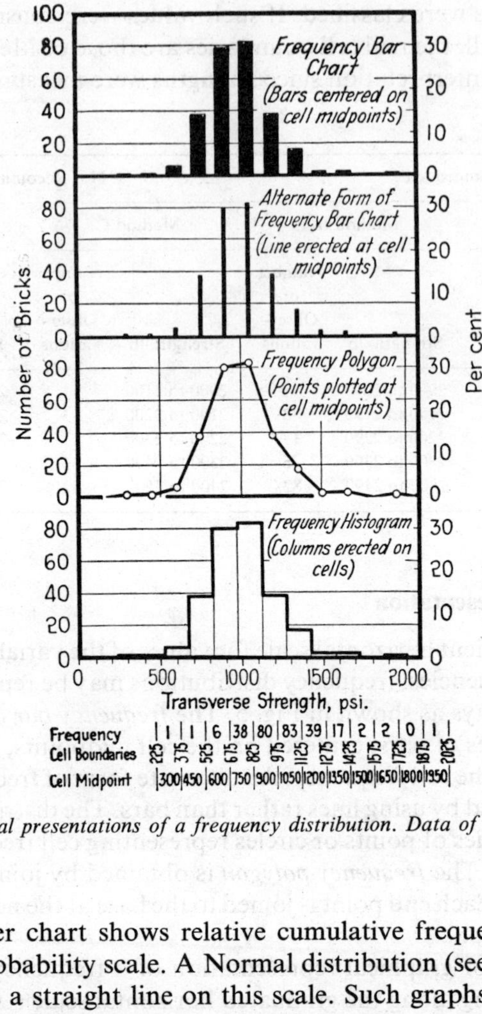

FIG. 5—*Graphical presentations of a frequency distribution. Data of Table 1*(a) *as grouped in Table 3*(a).

scale. The lower chart shows relative cumulative frequency plotted on a Normal law probability scale. A Normal distribution (see Fig. 13) will plot cumulatively as a straight line on this scale. Such graphs can be drawn to show the number of observations either "less than" or "greater than" the scale values. (Graph paper with one dimension graduated in terms of the summation of Normal law distribution has been described in Ref *3*).

15. Remarks

The information contained in the data may be summarized by presenting a tabular grouped frequency distribution, if the number of observations is large. A graphical presentation of a distribution makes it possible to visualize the nature and extent of the observed variation.

While some condensation is effected by presenting grouped frequency distributions, further reduction is necessary for most of the uses that are

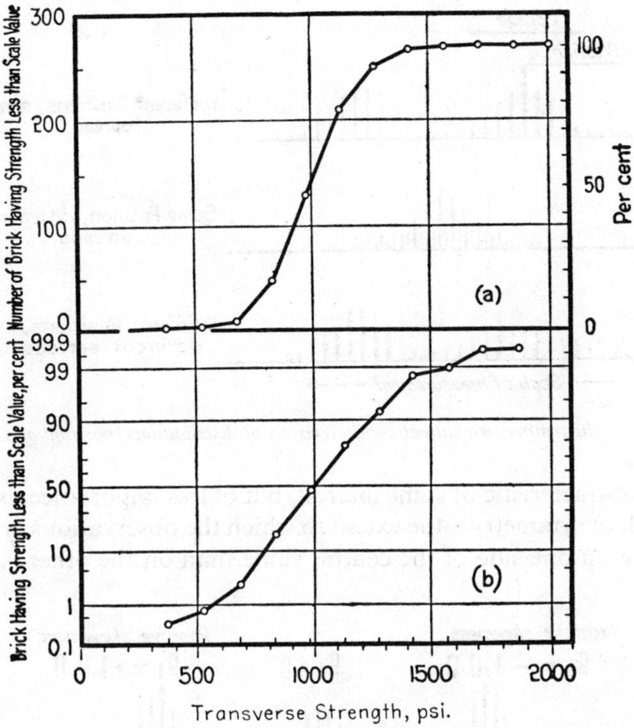

FIG. 6—*Graphical presentations of a cumulative frequency distribution. Data of Table 4:* (a) *using arithmetic scale for frequency, and* (b) *using probability scale for relative frequency.*

made of ASTM data. This need can be fulfilled by means of a few simple functions of the observed distribution, notably, the *average* and the *standard deviation.*

FUNCTIONS OF A FREQUENCY DISTRIBUTION

16. Introduction

In the problem of condensing and summarizing the information contained in the frequency distribution of a sample of observations, certain functions of the distribution are useful. For some purposes, a statement of the relative frequency within stated limits is all that is needed. For most purposes, however, two salient characteristics of the distribution which are illustrated in Fig. 7 are: (*a*) the position on the scale of measurement—the value about which the observations have a tendency to center, and (*b*) the spread or dispersion of the observations about the central value.

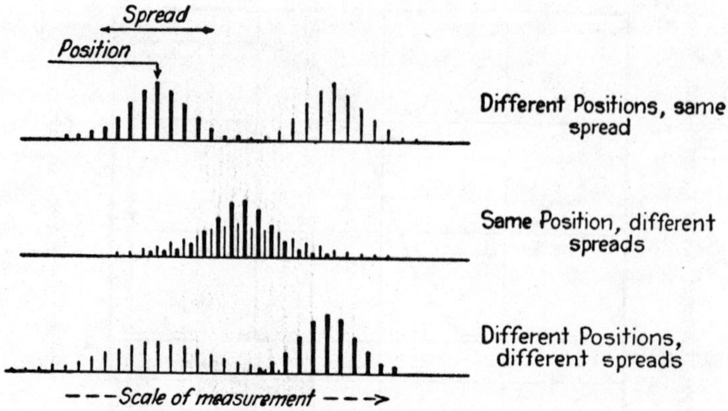

FIG. 7—*Illustrating two salient characteristics of distributions, position and spread.*

A third characteristic of some interest, but of less importance, is the skewness or lack of symmetry—the extent to which the observations group themselves more on one side of the central value than on the other (see Fig. 8).

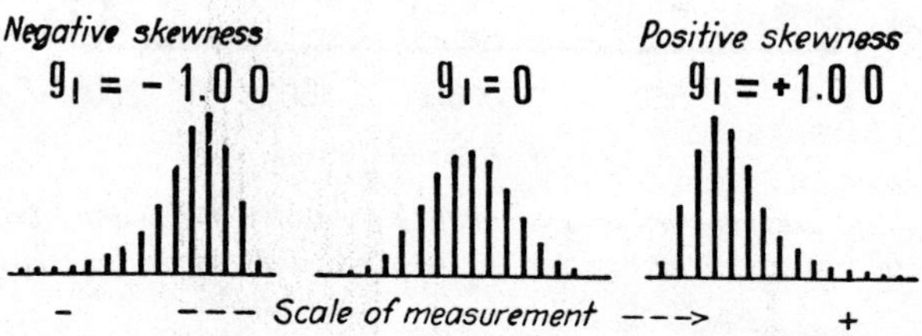

FIG. 8—*Illustrating a third characteristic of frequency distributions, skewness, and particular values of skewness,* g_1.

A fourth characteristic is "kurtosis" which relates to the tendency for a distribution to have a sharp peak in the middle and excessive frequencies on the tails as compared with the Normal distribution or conversely to be relatively flat in the middle with little or no tails (see Fig. 9).

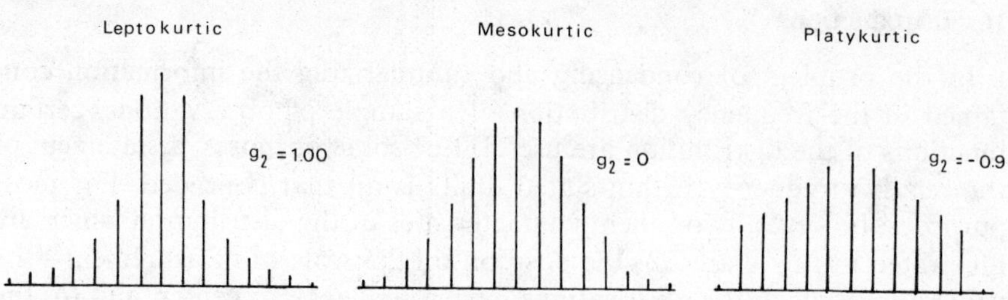

FIG. 9—*Illustrating the kurtosis of a frequency distribution and particular values of* g_2.

Several representative sample measures are available for describing these characteristics, but by far the most useful are the arithmetic mean, $\bar{X}$, the standard deviation, s, the skewness factor, g_1, and the kurtosis factor, g_2—all algebraic functions of the observed values. Once the numerical values of these particular measures have been determined, the original data may usually be dispensed with and two or more of these values presented instead.

17. Relative Frequency

The relative frequency, p, within stated limits on the scale of measurement is the ratio of the number of observations lying within those limits to the total number of observations.

In practical work, this function has its greatest usefulness as a measure of *fraction defective* or *fraction nonconforming*, in which case it is the fraction, p, representing the ratio of the number of observations lying outside specified limits (or beyond a specified limit) to the total number of observations.

18. Average (Arithmetic Mean)

The average (arithmetic mean) is the most widely used measure of central tendency. The term *average* and the symbol $\bar{X}$ will be used in this Manual to represent the arithmetic mean of a sample of numbers.

The average, $\bar{X}$, of a sample of n numbers, $X_1, X_2, \ldots X_n$, is the sum of the numbers divided by n, that is

$$\bar{X} = \frac{X_1 + X_2 + \cdots + X_n}{n} = \frac{\sum_{i=1}^{n} X_i}{n} \tag{1}$$

where the expression $\sum_{i=1}^{n} X_i$ means "the sum of all values of X, from X_1 to X_n, inclusive."

Considering the n values of X as specifying the positions on a straight line of n particles of equal weight, the average corresponds to the center of gravity of the system. The average of a series of observations is expressed in the same units of measurement as the observations, that is, if the observations are in pounds, the average is in pounds.

19. Other Measures of Central Tendency

The *geometric mean*, of a sample of n numbers, $X_1, X_2, \ldots X_n$, is the nth root of their product, that is

$$\text{geometric mean} = \sqrt[n]{X_1 X_2 \cdots X_n} \tag{2}$$

or

$$\log(\text{geometric mean}) = \frac{\log X_1 + \log X_2 + \cdots + \log X_n}{n} \tag{3}$$

Equation 3, obtained by taking logarithms of both sides of Eq 2, provides a convenient method for computing the geometric mean using the logarithms of the numbers.

NOTE—The distribution of some quality characteristics is such that a transformation, using logarithms of the observed values, gives a substantially Normal distribution. When this is true, the transformation is distinctly advantageous for (in accordance with Section 32) much of the total information can be presented by two functions, the average, $\overline{X}$, and the standard deviation, s, of the logarithms of the observed values. The problem of transformation is, however, a complex one that is beyond the scope of this Manual.

The *median* of the frequency distribution of n numbers is the middlemost value.

The *mode* of the frequency distribution of n numbers is the value which occurs most frequently.

20. Standard Deviation[3]

The standard deviation is the most useful measure of dispersion for the problems considered in Part 1 of the Manual.

For a sample of n numbers, $X_1, X_2, \ldots, X_n$, the sample standard deviation is commonly defined by the formula

$$s = \sqrt{\frac{(X_1 - \overline{X})^2 + (X_2 - \overline{X})^2 + \cdots + (X_n - \overline{X})^2}{n-1}} = \sqrt{\frac{\sum_{i=1}^{n} (X_i - \overline{X})^2}{n-1}} \tag{4}$$

where $\overline{X}$ is defined by Eq 1.

The quantity s^2 is called the sample *variance.*

The standard deviation of any series of observations is expressed in the same units of measurement as the observations, that is, if the observations are in pounds, the standard deviation is in pounds. (Variances would be measured in pounds squared.)

A frequently more convenient formula for the computation of s is

$$s = \sqrt{\frac{n \sum_{i=1}^{n} X_i^2 - \left(\sum_{i=1}^{n} X_i\right)^2}{n(n-1)}} \tag{5}$$

[3] Section 20 was rewritten in 1975.

A useful quantity related to the standard deviation is the root-mean-square deviation $\sigma = \sqrt{\sum (X - \overline{X})^2/n} = s\sqrt{(n-1)/n}$. This is used in formulas for Eqs 9 and 10 and in the discussion of Tchebychev's inequality in Section 32.

21. Other Measures of Dispersion

The *coefficient of variation, v*, of a sample of *n* numbers, is the ratio (sometimes the coefficient is expressed as a percentage) of their standard deviation, *s*, to their average, $\overline{X}$. It is given by

$$v = \frac{s}{\overline{X}} \tag{6}$$

The coefficient of variation is an adaptation of the standard deviation which was developed by Prof. Karl Pearson to express the variability of a set of numbers on a relative scale rather than on an absolute scale. It is thus a pure number.

The *average deviation* of a sample of *n* numbers, $X_1, X_2, \cdots X_n$, is the average of the absolute values of the deviations of the numbers from their average, $\overline{X}$, that is

$$\text{average deviation} = \frac{\sum_{i=1}^{n} |X_i - \overline{X}|}{n} \tag{7}$$

where the symbol $\|$ denotes the absolute value of the quantity enclosed.

The *range, R*, of a sample of *n* numbers is the difference between the largest number and the smallest number of the sample. This is the simplest measure of dispersion of a sample of observations.

22. Skewness—g_1

A useful measure of the lopsidedness of a sample frequency distribution is the coefficient of skewness g_1.

The coefficient of skewness g_1, of a sample of *n* numbers, $X_1, X_2, \cdots X_n$, is defined by the expression

$$g_1 = \frac{\sum_{i=1}^{n} (X_i - \overline{X})^3}{n\sigma^3} \tag{8}$$

This measure of skewness is a pure number and may be either positive or negative. For a symmetrical distribution, g_1 is zero. In general, for a non-symmetrical distribution, g_1 is negative if the long tail of the distribution extends to the left, the negative direction on the scale of measurement, and is positive if the long tail extends to the right, the positive direction on the scale of measurement. Figure 8 shows three unimodal distributions with different values of g_1.

22a. Kurtosis—g_2

The peakedness and tail excess of a sample frequency distribution is generally measured by the coefficient of kurtosis, g_2.

The coefficient of kurtosis, g_2, for a sample of n numbers, $X_1, X_2, \ldots, X_n$ is defined by the expression

$$g_2 = \frac{\sum_{i=1}^{n} (X_i - \overline{X})^4}{n\sigma^2} - 3 \qquad (9)$$

Again this is a pure number and may be either positive or negative. Generally, when a distribution has a sharp peak, thin shoulders, and fat tails relative to the bell-shaped distribution characterized by the Normal distribution, g_2 is positive. When a distribution is flat-topped with small tails, relative to the Normal distribution, g_2 is negative. Inverse relationships do not necessarily follow.[4] A distribution with a positive g_2 is said to be "leptokurtic." One with a negative g_2 is said to be "platykurtic." A distribution with $g_2 = 0$ is said to be "mesokurtic." Figure 9 gives three unimodal distributions with different values of g_2.

23. Remarks

Of the many measures that are available for describing the salient characteristics of a sample frequency distribution, the average, $\overline{X}$, the standard deviation, s, the skewness, g_1, and the kurtosis, g_2, are particularly useful for summarizing the information contained therein.

METHODS OF COMPUTING $\overline{X}$, s, g_1, AND g_2

24. Computation of Average and Standard Deviation

The average and standard deviation of a sample can be computed by using Eqs 1 and 5. The method of computation is illustrated in Table 5, using the data of Table 1 (*c*). The table of squares given in the Appendix is

[4] We cannot definitely infer anything about the shape of a distribution from knowledge of g_2 unless we are willing to assume some theoretical curve, say a Pearson curve, as being appropriate as a graduation formula (see Fig. 14 and Section 33).

TABLE 5—*Computation of average and standard deviation of a sample (data of Table 1 (c)).*

Specimen	Breaking Strength, X, lb	Square of Breaking Strength, X^2
1	578	334 084
2	572	327 184
3	570	324 900
4	568	322 624
5	572	327 184
6	570	324 900
7	570	324 900
8	572	327 184
9	576	331 776
10	584	341 056
$n = 10$	$\sum X = 5732$	$\sum X^2 = 3\ 285\ 792$

NOTE

sample *average*

$$\bar{X} = \frac{\Sigma X}{n} = \frac{5732}{10} = 573.2 \text{ lb}$$

sample standard deviation (using Eq 5)

$$s = \sqrt{\frac{n\Sigma X^2 - (\Sigma X)^2}{n(n-1)}}$$

$$= \sqrt{\frac{10(3\ 285\ 792) - (5732)^2}{10(9)}}$$

$$= \sqrt{23.29} = 4.83 \text{ lb}$$

useful in carrying out these computations. Square roots may be found to a satisfactory degree of accuracy for most practical purposes by using the columns of squares and interpolating to find the square roots desired.

The standard deviation of any set of numbers remains the same if a constant is added to or subtracted from each number in the set. Table 6 illustrates how this fact can be utilized to reduce the magnitude of the numbers dealt with in making computations. Subtraction of the constant A is equivalent to making computations with respect to an arbitrary origin, A. The computation work can readily be checked by using a second value of A.

Dividing or multiplying each of a set of numbers by a constant, has the effect of dividing or multiplying their standard deviation by that constant.

TABLE 6—*Illustrating simplified method of computing average and standard deviation of a sample.*

Original Data		Subtract $A = 5000$		Subtract $A = 5600$		Subtract $A = 5600$ and Divide by $h = 10$	
Original Number, X	Square, X^2	Number Minus A, X	Square, X^2	Number Minus A, X	Square, X^2	Number Minus A, Divided by h, X	Square, X^2
5 690	32 376 100	690	476 100	90	8 100	9	81
5 670	32 148 900	670	448 900	70	4 900	7	49
5 670	32 148 900	670	448 900	70	4 900	7	49
5 680	32 262 400	680	462 400	80	6 400	8	64
5 620	31 584 400	620	384 400	20	400	2	4
5 630	31 696 900	630	396 900	30	900	3	9
5 670	32 148 900	670	448 900	70	4 900	7	49
5 670	32 148 900	670	448 900	70	4 900	7	49
5 690	32 376 100	690	476 100	90	8 100	9	81
5 680	32 262 400	680	462 400	80	6 400	8	64

Computation Steps								
(*a*) total first column = ΣX	56 670	...	6 670	...	670	...	67	...
(*b*) total second column = ΣX^2	...	321 153 900	...	4 453 900	...	49 900	...	499
(*c*) divide (*a*) by 10 = $\Sigma X/10$	5 667	...	667	...	67	...	6.7	...
(*d*) add *A* to (*c*) = $A + \Sigma X/10$	$\bar{X} = 5\ 667$	...	$\bar{X} = 5\ 667$	...	$\bar{X} = 5\ 667$	...	$\bar{X} = 5\ 667$[a]	...
(*e*) multiply (*b*) by 10 = $10\Sigma X^2$	...	3 211 539 000	...	44 539 000	...	449 000	...	4 990
(*f*) square (*a*) = $(\Sigma X)^2$	...	3 211 488 900	...	44 488 900	...	448 900	...	4 489
(*g*) subtract (*f*) from (*e*) = $10\Sigma X^2 - (\Sigma X)^2$	...	50 100	...	50 100	...	50 100	...	501
(*i*) divide (*g*) by 10(9) = $[10\Sigma X^2 - (\Sigma X)^2]/10(9)$	...	556	...	556	...	556	...	5.56
(*j*) extract square root of (*i*) $= \sqrt{[10\Sigma X^2 - (\Sigma X)^2]/10(9)}$	...	$s = 23.6$	...	$s = 23.6$	...	$s = 23.6$	...	2.36
(*k*) multiply (*j*) by $h = 10$	...	...	...	...	...	...	...	$s = 23.6$

[a] Multiply (*c*) by $h = 10$ and add *A*.

The last two columns of Table 6 indicate how the arithmetic may be further simplified by dividing the original numbers by a constant, h.

25. Short Method of Computation When *n* is Large

When the number, n, of observations is large, the computation work can be simplified considerably by making use of the grouped frequency distribution of the observations. Table 7 shows a convenient form to use in computing the sample average, standard deviation, skewness, and kurtosis. With this method referred to as the "short method," an arbitrary origin, A, is used and deviations from this origin are expressed in cells rather than in units of the scale of measurement. The equations for the average, $\bar{X}$, the standard deviation, s, the skewness, g_1, and the kurtosis, g_2, of a grouped frequency are

$$\bar{X} = A + m\frac{\Sigma fd}{n} \tag{10}$$

$$s = m\sqrt{\frac{n\Sigma fd^2 - (\Sigma fd)^2}{n(n-1)}} \tag{11}$$

$$g_1 = \frac{n^3\Sigma fd^3 - 3n(\Sigma fd^2)(\Sigma fd) + 2(\Sigma fd)^2}{[n\Sigma fd^2 - (\Sigma fd)^2]^{3/2}} \tag{12}$$

$$g_2 = \frac{n^3\Sigma fd^4 - 4n^2(\Sigma fd^3)(\Sigma fd) + 6n(\Sigma fd^2)(\Sigma fd)^2 - 3(\Sigma fd)^4}{[n\Sigma fd^2 - (\Sigma fd)^2]^2} - 3 \tag{13}$$

where

A = arbitrary origin,
m = cell interval (difference between upper and lower boundaries of a cell),
f = observed cell frequency, and
d = deviation in cells from A.

Table 7 shows the computations for the data given in Table 1 (*a*). As will be noted, the work is simplified by making use of computation factors that are expressed in terms of the column totals, Σf, Σfd, Σfd^2, Σfd^3 and Σfd (see footnote 4 and Ref *1*, pp. 78–79).

The short methods of Table 7 are only applicable when the cell intervals are equal.

26. Remarks

The exact values of $\bar{X}$, s, g_1 and g_2 can, of course, be found by using Eqs 1, 5, 8, and 9, but the computation work may require an excessive amount of time when the number of observations is large. The short methods of computation (Section 25) introduce certain errors of grouping, since they assume that all observations in each cell have a value equal to that of the cell midpoint. It is believed, however, that the short methods are satisfactory for most practical purposes and that the errors introduced by grouping are not, in general, of sufficient importance to warrant the use of correction factors.

For the data of Table 1 (*a*), the errors introduced by grouping are indicated in the following tabulation.

	Exact Value	Value Found by Short Methods
Average, $\overline{X}$	1000.2	1000.0
Standard deviation, s	201.6	202.5

In carrying out the calculations, it is generally advisable to retain more places of figures throughout the work than are needed for final results, and throw away unneeded places only after the calculation work is completed (see Part 2, Section 7).

Programs for electronic computers are generally available for calculating $\overline{X}$, s, g_1, and g_2 or related quantities. Such programs should always be checked, however, to see just how the computations are carried out.

AMOUNT OF INFORMATION CONTAINED IN p, $\overline{X}$, s, g_1, AND g_2

27. Introduction

In this and following sections, the total information contained in a series of observations of a single variable is defined, and consideration is given to how much of the total information may be made available by presenting a few simple functions of the data—disregarding for the moment what uses are to be made of the data. The material in Sections 27 to 36 was revised in 1975 by introducing discussion of the kurtosis factor, g_2, and the use of Pearson frequency curves to graduate a distribution.

For present purposes, the *total information* will be defined as that contained in the original set of numbers arranged in ascending order of magnitude, that is, the ungrouped frequency distribution. (See Table 2.)

The concept of the ungrouped frequency distribution as giving the total information is set forth by Shewhart (Ref *1*, Chapter VIII). Since, in engineering practice, samples may not lightly be assumed to be random samples, additional information of value may be disclosed by considering the *order* of the observations.

TABLE 7—*Form for computing $\overline{X}$, s, g_1, and g_2 by short method (data of Table 1 (a)).*

Cell	Cell Midpoint	Cell Boundary	Deviation in Cells from A, d	Observed Frequency, f	fd	fd^2	fd^3	fd^4	Relative Frequency
0	A 300	225	0	1	0	0	0	0	0.4
1	450	375	1	1	1	1	1	1	0.4
2	600	525	2	6	12	24	48	96	2.2
3	750	675	3	38	114	342	1 026	3 078	14.1
4	900	825	4	80	320	1 280	5 120	20 480	29.6
5	1 050	975	5	83	415	2 075	10 375	51 875	30.7
6	1 200	1 125	6	39	234	1 404	8 424	50 544	14.5
7	1 350	1 275	7	17	119	833	5 831	40 817	6.3
8	1 500	1 425	8	2	16	128	1 024	8 192	0.7
9	1 650	1 575	9	2	18	162	1 458	13 122	0.7
10	1 800	1 725	10	0	0	0	0	0	0.0
11	1 950	1 875	11	1	11	121	1 331	14 641	0.4
12		2 025							
13									
14									
Total				n 270	Σfd 1 260	Σfd^2 6 370	Σfd^3 34 638	Σfd^4 202 846	100.0

Computations

NOTE—A = midpoint of Cell 0 = 300; m = cell interval = 150. The computation figures are carried out to a large number of places to prevent possible round-off errors.

$$\bar{X} = A + m\Sigma fd/n = 300 + 150(1\,260)/270 = 1\,000.0$$

$$s = m\sqrt{\frac{n\Sigma fd^2 - (\Sigma fd)^2}{n(n-1)}} = 150\sqrt{\frac{270(6\,370) - (1\,260)^2}{270(269)}} = 150\sqrt{\frac{1\,719\,900 - 1\,587\,600}{72\,630}} = 150\sqrt{1.821561} = 150(1.3497) = 202.5 \quad \sigma = s\sqrt{\frac{n-1}{n}} = 202.1$$

$$g_1 = \frac{n^2\Sigma fd^3 - 3n(\Sigma fd^2)(\Sigma fd) + 2(\Sigma fd)^3}{[n\Sigma fd^2 - (\Sigma fd)^2]^{3/2}} = \frac{(270)^2(34\,638) - 3(270)(6\,370)(1\,260) + 2(1\,260)^3}{\sqrt{(270(6\,370) - (1\,260)^2)^3}}$$

$$= \frac{2\,525\,110\,200 - 6\,501\,222\,000 + 4\,000\,752\,000}{\sqrt{(132\,300)^3}} = \frac{24\,640\,200}{48\,121\,297} = 0.51$$

$$g_2 = \frac{n^3\Sigma fd^4 - 4n^2(\Sigma fd^3)(\Sigma fd) + 6n(\Sigma fd^2)(\Sigma fd)^2 - 3(\Sigma fd)^4}{[n\Sigma fd^2 - (\Sigma fd)^2]^2} - 3 = \frac{(270)^3(202\,246) - 4(270)^2(34\,638)(1\,260) + 6(270)(6\,370)(1\,260)^2 - 3(1\,260)^4}{[270(6\,370) - (1\,260)^2]^2} - 3$$

$$= \frac{3\,980\,808\,018\,000 - 12\,726\,555\,408\,000 + 16\,383\,079\,440\,000 - 7\,561\,421\,280\,000}{17\,503\,290\,000} - 3 = \frac{75\,910\,770\,000}{17\,503\,290\,000} - 3 = 4.34 - 3 = 1.34$$

28. The Problem

Given a sample of n observations, $X_1, X_2, X_3, \ldots X_n$, of some quality characteristic, how can we present concisely information by means of which the observed distribution can be closely approximated, that is, so that the percentage of the total number, n, of observations lying within any stated interval from, say, $X = a$ to $X = b$, can be approximated?

The total information can be presented only by giving all of the observed values. It will be shown, however, that much of the total information is contained in a few simple functions—notably the average, $\bar{X}$, the standard deviation, s, the skewness, g_1, and the kurtosis, g_2.

29. Several Values of Relative Frequency, *p*

By presenting, say, 10 to 20 values of relative frequency, p, corresponding to stated cell intervals and also the number, n, of observations, it is possible to give practically all of the total information in the form of a tabular grouped frequency distribution. If the ungrouped distribution has any pecularities, however, the choice of cells may have an important bearing on the amount of information lost by grouping.

30. Single Value of Relative Frequency, *p*

If we present but a single value of relative frequency, p, such as the fraction of the total number of observed values falling outside of a specified limit and also the number, n, of observations, the portion of the total information presented is very small. This follows from the fact that quite dissimilar distributions may have identically the same value of p as illustrated in Fig. 10.

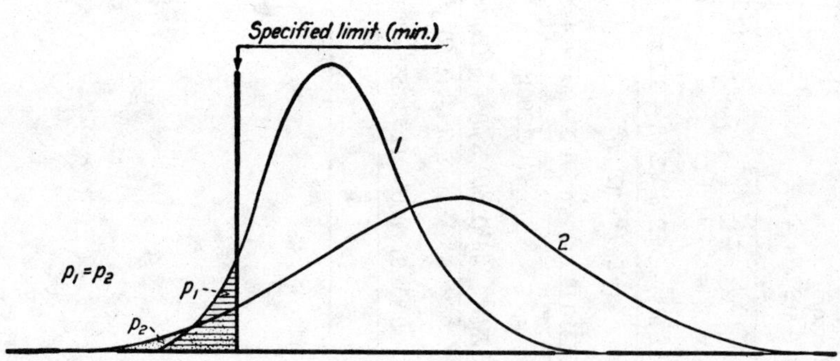

FIG. 10—*Quite different distributions may have the same value of* p, *fraction of total observations below specified limit.*

NOTE—For the purposes of Part 1 of the Manual, the curves of Figs. 10 and 11 may be taken to represent frequency histograms with small cell widths and based on large samples. In a frequency histogram, such as that shown at the bottom of Fig. 5, let the percentage relative frequency between any two cell boundaries be represented by the *area* of the histogram between those boundaries, the total area being 100 percent. Since the cells are of uniform width, the relative frequency in any cell is then proportional to the *height* of that cell and may be read on the vertical scale to the right.

If the sample size is increased and the cell width reduced, a histogram in which the relative frequency is measured by area approaches as a limit the frequency distribution of the universe, which in many cases can be represented by a smooth curve. The relative frequency between any two values is then represented by the *area* under the curve and between ordinates erected at those values. Because of the method of generation, the ordinate of the curve may be regarded as a curve of *relative frequency density*. This is analogous to the representation of the variation of density along a rod of uniform cross section by a smooth curve. The weight between any two points along the rod is proportional to the area under the curve between the two ordinates and we may speak of the *density* (that is, weight density) at any point but not of the *weight* frequency at any point.

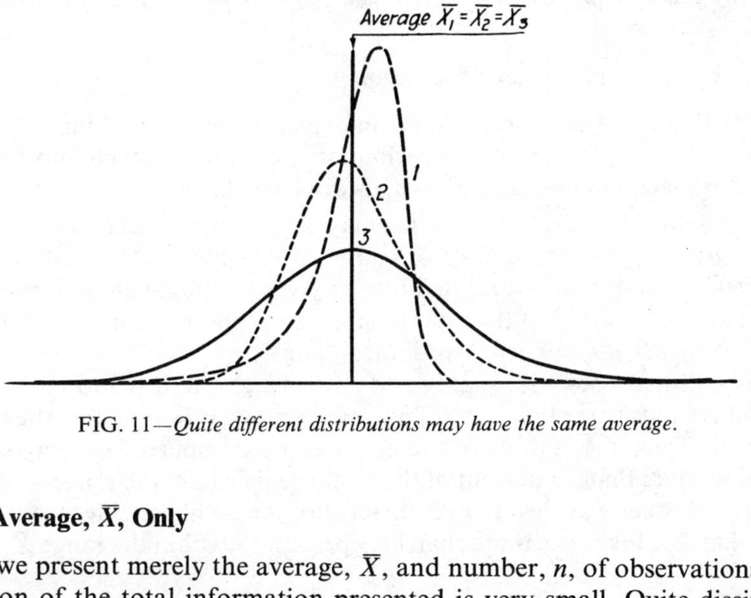

FIG. 11—*Quite different distributions may have the same average.*

31. Average, $\bar{X}$, Only

If we present merely the average, $\bar{X}$, and number, n, of observations, the portion of the total information presented is very small. Quite dissimilar distributions may have identically the same value of $\bar{X}$, as illustrated in Fig. 11.

In fact, no single one of the five functions, p, $\bar{X}$, s, g_1, or g_2 presented alone is generally capable of giving much of the total information in the original distribution. Only by presenting two or three of these functions can a fairly complete description of the distribution generally be made.

An exception to the above statement occurs when theory and observation suggest that the underlying law of variation is a distribution for which the basic characteristics are all functions of the mean. For example, "life" data "under controlled conditions" sometimes follow a negative exponential distribution. (For a more general form for the distribution of the life of a product, see Supplement, p. 46.) For this the cumulative relative frequency is given by the equation

$$F(X) = 1 - e^{-X/\theta} \qquad 0 < X < \infty \tag{14}$$

This is a single parameter distribution for which the mean and standard deviation both equal θ. That the negative exponential distribution is the underlying law of variation can be checked by noting whether values of $1 - F(X)$ for the sample data tend to plot as a straight line on ordinary 1-cycle logarithmic paper. In such a situation knowledge of $\bar{X}$ will, by taking $\theta = \bar{X}$ in Eq 14 and using tables of the exponential function, yield a graduating formula from which estimates can be made of the percentage of cases lying between any two specified values of X. Presentation of $\bar{X}$ and n is sufficient in such cases provided they are accompanied by a statement that there are reasons to believe that X has a negative exponential distribution.

32. Average, $\bar{X}$, and Standard Deviation, s

These two functions contain some information even if nothing is known about the form of the observed distribution, and contain much information when certain conditions are satisfied, as discussed later.

Noting that $\sigma = s\sqrt{(n-1)/n}$ we may say, with no reservations whatsoever, that the presentation of $\bar{X}$ and s, together with the number, n, of observations and the derived quantity, σ, gives the following information.

More than $1 - 1/k^2$ of the total number, n, of observations lie within the closed range $\bar{X} \pm k\sigma$ (where k is not less than 1).

This is Tchebychev's inequality and is shown graphically in Fig. 12. The inequality holds true of *any* set of finite numbers regardless of how they were obtained. Thus if $\bar{X}$ and s are presented and σ computed, we may say at once that more than 75 percent of the numbers lie within the range $\bar{X} \pm 2\sigma$; stated in another way, less than 25 percent of the numbers differ from $\bar{X}$ by more than 2σ. Likewise, more than 88.9 percent lie within the range $\bar{X} \pm 3\sigma$,

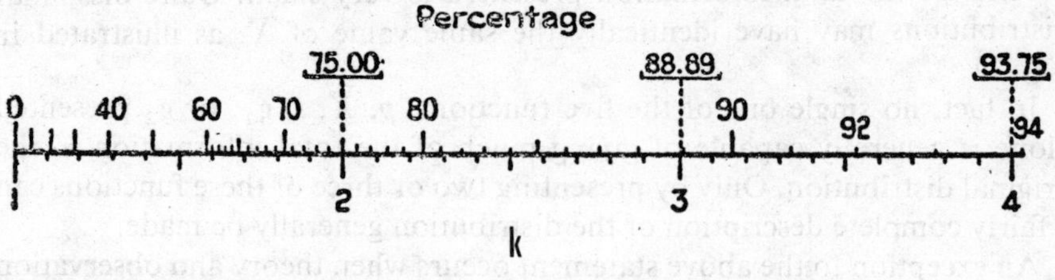

FIG. 12—*Percentage of the total observations lying within the range* X ± kσ *always exceeds the percentage given on this chart.*

etc. From this inequality we also have the rule that if $n = 4$, all observations fall within $\bar{X} \pm 2\sigma$, if $n = 10$, all observations fall within $\bar{X} \pm 3.16\sigma$, etc., as shown in Fig. 13. This rule is useful particularly when n is small. Table 8

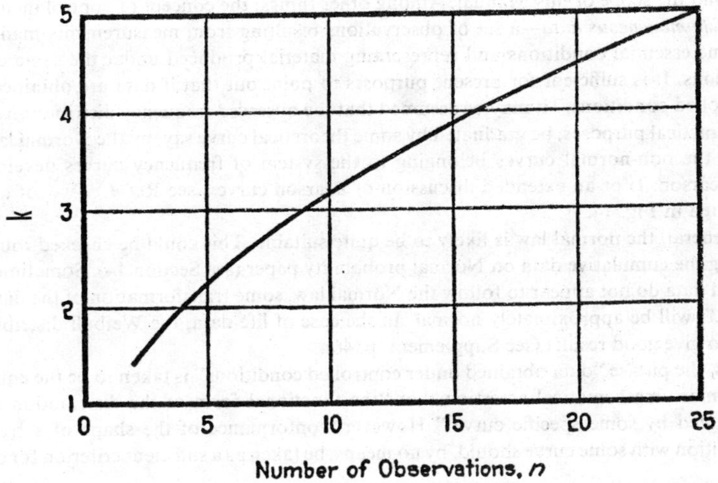

FIG. 13—*Values of* k[1], *such that* all *observations lie within the range* X ± kσ.

TABLE 8—*Comparison of observed percentages and theoretical minimum percentages of the total observations lying within given ranges.*

Range, $\bar{X} \pm k\sigma$	Theoretical Percentages of Total Observations Lying Within the Given Range $\bar{X} \pm k\sigma$	Observed Percentages[a]		
		Data of Table 1 (*a*) (n = 270)	Data of Table 1 (*b*) (n = 100)	Data of Table 1 (*c*) (n = 10)
$\bar{X} \pm 2.0\sigma$	more than 75.0	96.7	94	90
$\bar{X} \pm 2.5\sigma$	more than 84.0	97.8	100	90
$\bar{X} \pm 3.0\sigma$	more than 88.9	98.5	100	100

[a] Data of Table 1 (*a*). $\bar{X} = 1000$, $\sigma = 202$; data of Table 1 (*b*), $\bar{X} = 1.535$, $\sigma = 0.105$; data of Table 1 (*c*), $\bar{X} = 573.2$, $\sigma = 4.58$.

indicates the conformance with Tchebychev's inequality of the three sets of observations given in Table 1.

To determine approximately just what percentages of the total number of observations lie within given limits, as contrasted with minimum percentages within those limits (given above by Tchebychev's inequality), requires additional information of a restrictive nature. If we present $\bar{X}$, s, and n, and are able to add the information "data obtained under controlled conditions," then it is possible to make such estimates satisfactorily for limits spaced equally above and below $\overline{X}$.

NOTE—What is meant technically by "controlled conditions" is discussed by Shewhart [*1*] and is beyond the scope of this Manual. Among other things, the concept of control includes the idea of *homogeneous data*—a set of observations resulting from measurements made under the same essential conditions and representing material produced under the same essential conditions. It is sufficient for present purposes to point out that if data are obtained under "controlled conditions," it may be assumed that the observed frequency distribution can, for most practical purposes, be graduated by some theoretical curve say, by the Normal law or by one of the non-normal curves belonging to the system of frequency curves developed by Karl Pearson. (For an extended discussion of Pearson curves, see Ref *4*.) Two of these are illustrated in Fig. 14.

In general, the normal law is likely to be quite suitable. This could be checked roughly by plotting the cumulative data on Normal probability paper (see Section 14). Sometimes if the original data do not appear to follow the Normal law, some transformation of the data, such as log X, will be approximately normal. In the case of life data, the Weibull distribution is likely to give good results (see Supplement p. 46).

Thus, the phrase "data obtained under controlled conditions" is taken to be the equivalent of the more mathematical assertion that "the functional form of the distribution may be represented by some specific curve." However, conformance of the shape of a frequency distribution with some curve should, by no means, be taken as a sufficient criterion for control.

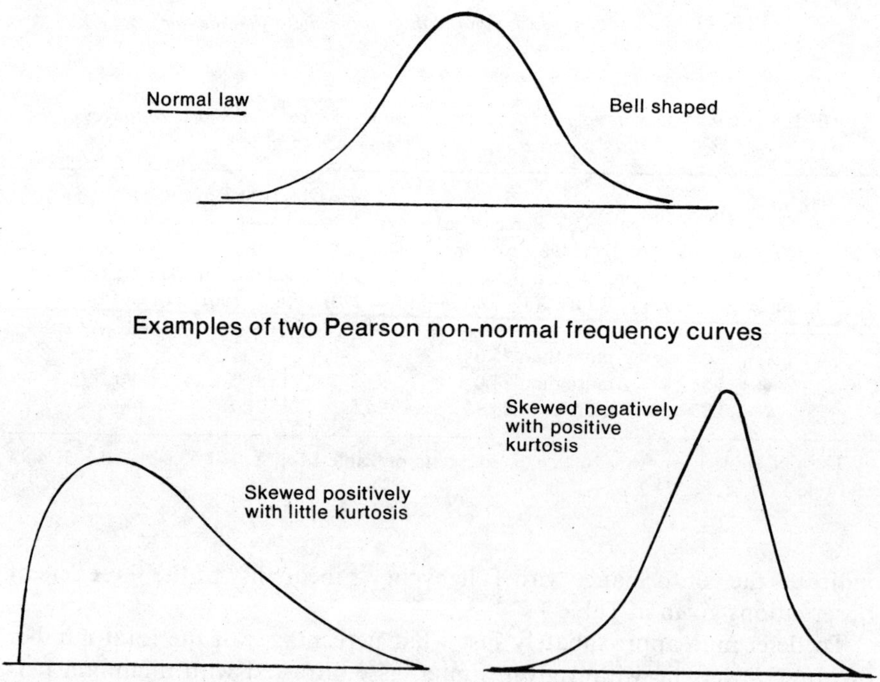

FIG. 14—*A frequency distribution of observations obtained under controlled conditions will usually have an outline that conforms with the Normal law or a non-normal Pearson frequency curve.*

Generally for controlled conditions, the percentage of the total observations in the original sample lying within the range $\bar{X} \pm k\sigma$ may be determined approximately from the chart of Fig. 15, which is based on the Normal law

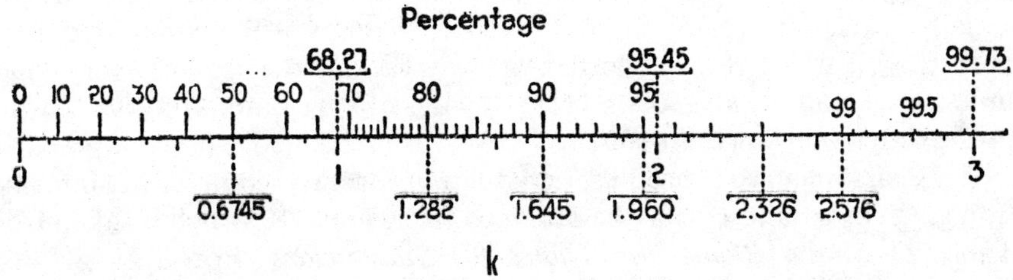

FIG. 15—*Normal law integral diagram giving percentage of total area under Normal law curve falling within the range* $\bar{X}' \pm k\sigma'$. *This diagram is also useful in probability and sampling problems, expressing the upper (percentage) scale values in decimals to represent "probability."*

integral. The approximation may be expected to be better the larger the number of observations. Table 9 compares the observed percentages of the total number of observations lying within several symmetrical ranges about $\bar{X}$ with those estimated from a knowledge of $\bar{X}$ and σ, for the three sets of observations given in Table 1.

TABLE 9—*Comparison of observed percentages and theoretical estimated percentages of the total observations lying within given ranges.*

Range, $\bar{X} \pm k\sigma$	Theoretical Estimated Percentages[a] of Total Observations Lying Within the Given Range $\bar{X} \pm k\sigma$	Observed Percentages		
		Data of Table 1 (*a*) (n = 270)	Data of Table 1 (*b*) (n = 100)	Data of Table 1 (*c*) (n = 10)
$\bar{X} \pm 0.6745\sigma$	50.0	52.2	54	70
$\bar{X} \pm 1.0\sigma$	68.3	76.3	72	80
$\bar{X} \pm 1.5\sigma$	86.6	89.3	84	90
$\bar{X} \pm 2.0\sigma$	95.5	96.7	94	90
$\bar{X} \pm 2.5\sigma$	98.7	97.8	100	90
$\bar{X} \pm 3.0\sigma$	99.7	98.5	100	100

[a] Use Fig. 15 with $\bar{X}$ and σ as estimates of $\bar{X}'$ and σ'.

33. Average, $\bar{X}$, Standard Deviation, s, Skewness, g_1, and Kurtosis, g_2

If the data are obtained under "controlled conditions" and if a Pearson curve is assumed to be appropriate as a graduation formula, the presentation of g_1 and g_2 in addition to $\overline{X}$ and s will contribute further information. They will give no immediate help in determining the percentage of the total observations lying within a symmetric range about the average $\overline{X}$, that is, in the range of $\overline{X} \pm k\sigma$. What they do is to help in estimating observed percentages (in a sample already taken) in a range whose limits are not equally spaced above and below $\overline{X}$. In the opinion of the committee, however, the g_1 and g_2 for single sets of observations do not help very much unless n is greater than 250, and for purposes of Part 1 are rarely worthwhile presenting if n is less than 100.

If a Pearson curve is used as a graduation formula, some of the information given by g_1 and g_2 may be obtained from Table 10 which is taken from Table 42 of the *Biometrika Tables for Statisticians*. For $\beta_1 = g_1{}^2$ and $\beta_2 = g_2 + 3$, this table gives values of k_L for use in estimating the lower 2.5 percent of the data and values of k_U for use in estimating the upper 2.5 percent point. More specifically, it may be estimated that 2.5 percent of the cases are less than $\overline{X} - k_L\sigma$ and 2.5 percent are greater than $\overline{X} + k_U\sigma$. Put another way, it may be estimated that 95 percent of the cases are between $\overline{X} - k_L\sigma$ and $\overline{X} + k_U\sigma$. Table 42 of the *Biometrika Tables for Statisticians* also give values of k_L and k_U for 0.5, 1.0, and 5.0 percent points.

Example

Table 7 computes $\overline{X}, s, \sigma, g_1$, and g_2 for a sample of 270 observations of the transverse strength of bricks. The sample distribution is shown in Fig. 4.

From the sample values of $g_1 = 0.51$ and $g_2 = 1.34$, we take $\beta_1 = g_1{}^2 = (0.51)^2 = 0.26$ and $\beta_2 = 1.34 + 3 = 4.34$. Thus from Tables 10 (*a*) and (*b*) we may estimate that approximately 95 percent of the 270 cases lie between $\overline{X} - k_L\sigma$ and $\overline{X} + k_U\sigma$ or between $1000 - 1.801\ (202.1) = 636.0$ and $1000 + 2.17\ (202.1) = 1438.6$. The actual percentage of the 270 cases in this range is 96.3 percent. (See Table 2 (*a*).) Presentation of the four statistics, $\overline{X}$, s, g_1, and g_2 is thus sufficient to enable us to determine with reasonable accuracy the range within which there is a high percentage of the cases.

NOTE ON TOLERANCE LIMITS—In Sections 32 through 33 the percentages of X values estimated to be within a specified range pertain only to the given sample of data which is being represented succinctly by selected statistics, $\overline{X}$, s, etc. The Pearson curves used to derive these percentages are used simply as graduation formulas for the histogram of the sample data. The aim of Sections 32 to 33 is to indicate how much information about the sample is given by $\overline{X}$, s, g_1 and g_2. *It should be carefully noted that in an analysis of this kind the selected ranges of X and associated percentages are not to be confused with what in the statistical literature are called "tolerance limits."*

In statistical analysis tolerance limits are values on the X scale that denote a range which may be stated to contain a specified minimum percentage of the values in the universe, there being attached to this statement a coefficient indicating the degree of confidence in its truth. For example, with reference to a random sample of 400 items, it may be said, with a 0.91 probability of being right, that 99 percent of the values in the universe from which the sample came will be in the range $X_H - X_L$ where X_H and X_L are respectively the largest and smallest values in the sample. If the universe is known to be Normal it might also be said, with a 0.90 probability of being right, that 99 percent of the values of the population will lie in the range $X \pm 2.703s$. Further information on statistical tolerances of this kind is presented in Refs 5–7.

34. Use of Coefficient of Variation Instead of the Standard Deviation

So far as quantity of information is concerned, the presentation of the sample coefficient of variation, v, together with the average, $\bar{X}$, is equivalent to presenting the sample standard deviation, s, and the average, $\bar{X}$, since s may be computed directly from the values of $v = s/\bar{X}$ and $\bar{X}$. In fact the sample coefficient of variation (multiplied by 100) is merely the sample standard deviation, s, expressed as a percentage of the average, $\bar{X}$. The coefficient of variation is sometimes useful in presentations whose purpose is to compare variabilities, relative to the averages, of two or more distributions.

Example 1

The following table presents strength test results for two different materials. It can be seen that whereas the standard deviation for Material B is less than the standard deviation for Material A, the latter shows the greater relative variability as measured by the coefficient of variability.

Material	Number of Observations, n	Average Strength, lb, $\bar{X}$	Standard Deviation, lb, s	Coefficient of Variation, %, v
A	160	1100	225	0.204
B	150	800	200	0.250

The coefficient of variation is particularly applicable in reporting the results of certain measurements where the variability, σ', is known or suspected to be a function of the level of the measurements. Such a situation may be encountered when it is desired to compare the variability (*a*) of physical properties of related materials usually at different levels, (*b*) of the performance of a material under two different test conditions, or (*c*) of analyses for a specific element or compound present in different concentrations.

Example 2

The performance of a material may be tested under widely different test conditions as for instance in a standard life test and in an accelerated life

TABLE 10—*Lower and upper 2.5 percent points of the standardized deviate* $(X - \mu)/\sigma'$, *given by Pearson frequency curves for designated values of* β_1 *(estimated as equal to* $g_1{}^2$*) and* β_2 *(estimated as equal to* $g_2 + 3$*).*

	β_1 / β_2	0.00	0.01	0.03	0.05	0.10	0.15	0.20	0.30	0.40	0.50	0.60	0.70	0.80	0.90	1.00
(a) Lower[a]	1.8	1.65	...	...	...	...	...	...	...	...	...	...	...	...	...	...
	2.0	1.76	1.68	1.62	1.56	...	...	...	...	...	...	...	...	...	...	...
	2.2	1.83	1.76	1.71	1.66	1.57	1.49	1.41	...	...	...	...	...	...	...	...
	2.4	1.88	1.82	1.77	1.73	1.65	1.58	1.51	1.39	...	...	...	...	...	...	...
	2.6	1.92	1.86	1.82	1.78	1.71	1.64	1.58	1.47	1.37	...	...	...	...	...	...
	2.8	1.94	1.89	1.85	1.82	1.76	1.70	1.65	1.55	1.45	1.35	...	...	...	...	...
	3.0	1.96	1.91	1.87	1.84	1.79	1.74	1.69	1.60	1.52	1.42	1.33	...	...	...	...
	3.2	1.97	1.93	1.89	1.86	1.81	1.77	1.72	1.65	1.57	1.49	1.40	1.32	1.24	...	...
	3.4	1.98	1.94	1.90	1.88	1.83	1.79	1.75	1.68	1.61	1.54	1.46	1.39	1.31	1.23	...
	3.6	1.99	1.95	1.91	1.89	1.85	1.81	1.77	1.71	1.65	1.58	1.51	1.44	1.38	1.30	1.23
	3.8	1.99	1.95	1.92	1.90	1.86	1.82	1.79	1.73	1.67	1.62	1.56	1.49	1.43	1.36	1.29
	4.0	1.99	1.96	1.93	1.91	1.87	1.84	1.81	1.75	1.70	1.64	1.59	1.53	1.47	1.41	1.35
	4.2	2.00	1.96	1.93	1.91	1.88	1.84	1.82	1.76	1.72	1.67	1.62	1.56	1.51	1.45	1.40
	4.4	2.00	1.96	1.94	1.92	1.88	1.85	1.83	1.78	1.73	1.69	1.64	1.59	1.54	1.49	1.44
	4.6	2.00	1.96	1.94	1.92	1.89	1.86	1.83	1.79	1.75	1.70	1.66	1.62	1.57	1.52	1.47
	4.8	2.00	1.97	1.94	1.93	1.89	1.87	1.84	1.80	1.76	1.72	1.68	1.64	1.59	1.55	1.50
	5.0	2.00	1.97	1.94	1.93	1.90	1.87	1.85	1.81	1.77	1.73	1.69	1.65	1.61	1.57	1.53

(b) Upper	1.8	1.65	...	...	...	...	...	...	...	...	...	...	...	...	...	...
	2.0	1.76	1.82	1.86	1.89	...	...	...	...	...	...	...	...	...	...	...
	2.2	1.83	1.89	1.93	1.96	2.00	2.04	2.06	...	...	...	...	...	...	...	...
	2.4	1.88	1.94	1.98	2.01	2.05	2.08	2.11	2.15	...	...	...	...	...	...	...
	2.6	1.92	1.97	2.01	2.03	2.08	2.11	2.14	2.18	2.22	...	...	...	...	...	...
	2.8	1.94	1.99	2.03	2.05	2.09	2.13	2.15	2.20	2.24	2.27	...	...	...	...	...
	3.0	1.96	2.01	2.04	2.06	2.10	2.13	2.16	2.21	2.25	2.28	2.32	...	...	...	...
	3.2	1.97	2.02	2.05	2.07	2.11	2.14	2.16	2.21	2.25	2.29	2.32	2.35	2.38	...	...
	3.4	1.98	2.02	2.05	2.07	2.11	2.14	2.16	2.21	2.25	2.28	2.32	2.35	2.38	2.41	...
	3.6	1.99	2.02	2.05	2.07	2.11	2.14	2.16	2.20	2.24	2.28	2.31	2.34	2.37	2.41	2.44
	3.8	1.99	2.03	2.05	2.07	2.11	2.13	2.16	2.20	2.24	2.27	2.30	2.33	2.36	2.40	2.43
	4.0	1.99	2.03	2.05	2.07	2.11	2.13	2.15	2.19	2.23	2.26	2.29	2.32	2.35	2.38	2.41
	4.2	2.00	2.03	2.05	2.07	2.10	2.13	2.15	2.19	2.22	2.25	2.28	2.31	2.34	2.37	2.40
	4.4	2.00	2.03	2.05	2.07	2.10	2.13	2.15	2.18	2.22	2.25	2.28	2.31	2.33	2.36	2.39
	4.6	2.00	2.03	2.05	2.07	2.10	2.12	2.14	2.18	2.21	2.24	2.27	2.30	2.32	2.35	2.38
	4.8	2.00	2.03	2.05	2.07	2.10	2.12	2.14	2.17	2.21	2.23	2.26	2.29	2.31	2.34	2.36
	5.0	2.00	2.03	2.05	2.07	2.09	2.12	2.14	2.17	2.20	2.23	2.25	2.28	2.30	2.33	2.35

NOTES—

This table was reproduced from *Biometrika Tables for Statisticians,* Vol. 1, p. 207, with the kind permission of the Biometrika Trust. The Biometrika Tables also give the lower and upper 0.5, 1.0, and 5 percent points. Use for a large sample only, say $n \geq 250$. Take $\mu = \bar{X}$ and $\sigma' = \sigma$.

[a]When $g_1 > 0$, the skewness is taken to be positive and the deviates for the lower percentage points are negative.

test. Further, the units of measurement of the accelerated life tester may be in minutes and of the standard tester, in hours. The following data indicate essentially the same relative variability of performance for the two test conditions.

Test Condition	Number of Specimens, n	Average Life, $\bar{X}$	Standard Deviation, s	Coefficient of Variation, % v
A	50	14 h	4.2 h	0.300
B	50	80 min	23.2 min	0.290

35. General Comment on Observed Frequency Distributions of a Series of ASTM Observations

Experience with frequency distributions for physical characteristics of materials and manufactured products prompts the committee to insert a comment at this point. We have yet to find an observed frequency distribution of over 100 observations of a quality characteristic and purporting to represent essentially uniform conditions, that has less than 96 percent of its values within the range $\bar{X} \pm 3\sigma$. For a Normal distribution, 99.7 percent of the cases lie between $\bar{X}' \pm 3\sigma'$, as indicated in Fig. 15. Taking this as a starting point and considering the fact that in ASTM work the intention is, in general, to avoid throwing together into a single series data obtained under widely different conditions—different in an important sense in respect to the characteristic under inquiry—we believe that it is possible, in general, to use the methods indicated in Sections 32 and 33 for making rough estimates of the observed percentages of a frequency distribution, at least for making estimates (per Section 32) for symmetric ranges around the average, that is, $\bar{X} \pm k\sigma$. This belief depends, to be sure, upon our own experience with frequency distributions and upon the observation that such distributions tend, in general, to be unimodal—to have a single peak—as in Fig. 14.

Discriminate use of these methods is, of course, presumed. The methods suggested for controlled conditions could not be expected to give satisfactory results if the parent distribution were one like that shown in Fig. 16—a bimodal distribution representing two different sets of conditions. Here, however, the methods could be applied separately to each of the two rational subgroups of data.

36. Summary—Amount of Information Contained in Simple Functions of the Data

The material given in Sections 27 to 35, inclusive, may be summarized as follows.

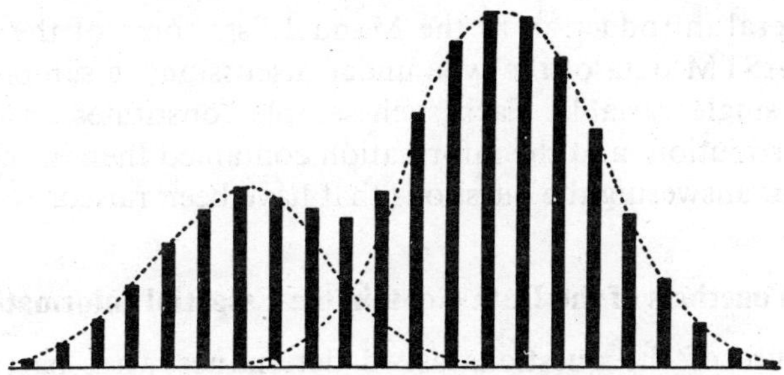

FIG. 16—*A bimodal distribution arising from two different systems of causes.*

1. If a sample of observations of a single variable is obtained under controlled conditions, much of the total information contained therein may be made available by presenting four functions—the average, $\overline{X}$, the standard deviation, s, the skewness, g_1, the kurtosis, g_2—and the number, n, of observations. Of the four functions, $\overline{X}$ and s contribute most; g_1 and g_2 contribute something but, in the opinion of the committee, not very much unless n is greater than, say, 250.

2. The average, $\overline{X}$, and the standard deviation, s, give some information even for data that are not obtained under controlled conditions.

3. No single function, such as the average, of a sample of observations is capable of giving much of the total information contained therein unless the sample is from a universe that is itself characterized by a single parameter.

Just what functions of the data should be presented in any instance depends on what uses are to be made of the data. This leads to a consideration of what constitutes the "essential information."

ESSENTIAL INFORMATION

37. Introduction

Presentation of data presumes some intended use either by others or by the author as supporting evidence for his conclusions. The objective is to present that portion of the total information given by the original data that is believed to be essential for the intended use. *Essential information* will be described as follows, taken from Ref *1*, p. 58.

> "We take data to answer specific questions. We shall say that a set of statistics (functions) for a given set of data contains the *essential information* given by the data when, through the use of these statistics, we can answer the questions in such a way that further analysis of the data will not modify our answers to a practical extent."

The general introduction to the Manual lists some of the objectives of gathering ASTM data of the type under discussion—a sample of observations of a single variable. Each such sample constitutes an observed frequency distribution, and the information contained therein should be used efficiently in answering the questions that have been raised.

38. What Functions of the Data Contain the Essential Information

The nature of the questions asked determines what part of the total information in the data constitutes the essential information for use in interpretation.

If we are interested in the percentages of the total number of observations that have values above (or below) several values on the scale of measurement, the essential information may be contained in a tabular grouped frequency distribution plus a statement of the number of observations, n. But even here, if n is large and if the data represent controlled conditions, the essential information may be contained in the four sample functions—the average, $\overline{X}$, the standard deviation, s, the skewness, g_1, the kurtosis, g_2—and the number of observations, n. If we are interested in the average and variability of the quality of a material, or in the average quality of a material and some measure of the variability of averages for successive samples, or in a comparison of the average and variability of the quality of one material with that of other materials, or in the error of measurement of a test, or the like, then the essential information may be contained in the $\overline{X}$, s and n of each sample of observations. Here, if n is small, say 10 or less, much of the essential information may be contained in the $\overline{X}$, R (range), and n of each sample of observations.

It is important to note [*8*] that the expected value of the range, R (largest observed value minus smallest observed value) for samples of n observations each, drawn from a Normal universe having a standard deviation σ', varies with sample size in the following manner.

The expected value of the range is 2.1 σ' for $n = 4$, 3.1 σ' for $n = 10$, 3.9 σ' for $n = 25$, and 6.1 σ' for $n = 500$. From this it is seen that in sampling from a Normal universe, the spread between the maximum and the minimum observation may be expected to be about twice as great for a sample of 25, and about three times as great for a sample of 500, as for a sample of 4. For this reason, n should *always* be given in presentations which give R. In general, it is better not to use R if $n > 12$.

If we are also interested in the percentage of the total quantity of product that does not conform with specified limits, then part of the essential information may be contained in the observed value of fraction defective, p.

If the conditions under which the data were obtained were not controlled, the maximum and minimum observations may contain information of value.

It is to be carefully noted that if our interest goes beyond the sample data themselves to the processes that generated the samples or might generate similar samples in the future, we need to consider errors that may arise from sampling. The problems of sampling errors that arise in estimating process means, variances, and percentages are discussed in Part 2 that follows. For discussions of sampling errors in comparisons of means and variabilities of different samples, the reader is referred to texts on statistical theory (for example, Ref *9*). The intention here is simply to note those statistics, those functions of the sample data, that would be useful in making such comparisons and consequently should be reported in the presentation of sample data.

39. Presenting $\bar{X}$ Only Versus Presenting $\bar{X}$ and s

Presentation of the essential information contained in a sample of observations commonly consists in presenting $\bar{X}$, s, and n. Sometimes the average alone is given—no record is made of the dispersion of the observed values nor of the number of observations taken. For example, Table 11 gives the

TABLE 11—*Information of value may be lost if only the average is presented.*

	Tensile Strength, psi		
Material	Condition *a* Average, $\bar{X}$	Condition *b* Average, $\bar{X}$	Condition *c* Average, $\bar{X}$
A	51 430	47 200	49 010
B	59 060	57 380	60 700
C	57 710	74 920	80 460

observed average tensile strength for several materials under several conditions. The objective quality in each instance is a frequency distribution, from which the set of observed values may be considered as a sample. Much information of value is generally lost by presenting merely the average, and failing to present some measure of dispersion and the number of observations.

Table 12 corresponds to Table 11 and provides what will usually be considered as the essential information for several sets of observations, such as are obtained in investigations conducted for the purpose of comparing the quality of different materials.

40. Observed Relationships

ASTM work often requires the presentation of data showing the observed relationship between two variables. Although strictly this subject does not

TABLE 12—*Presentation of essential information (data of Table 11).*

		Tensile Strength, psi					
		Condition *a*		Condition *b*		Condition *c*	
Material	Tests	Average, $\bar{X}$	Standard Deviation, s	Average, $\bar{X}$	Standard Deviation, s	Average, $\bar{X}$	Standard Deviation, s
A	20	51 430	920	47 200	830	49 010	1070
B	18	59 060	1320	57 380	1 360	60 700	1480
C	27	75 710	1840	74 920	1 650	80 460	1910

fall within the scope of Part 1 of the Manual, the following material is included for general information.

Attention will be given here to one type of relationship, where one of the two variables is of the nature of temperature or time—one that is controlled at will by the investigator and considered for all practical purposes as capable of "exact" measurement, free from experimental errors. (The problem of presenting information on the observed relationship between *two statistical variables*, such as hardness and tensile strength of an alloy sheet material, is more complex and will not be treated here. For further information, see Refs *1*, *2*, and *9*.) Such relationships are commonly presented in the form of a chart consisting of a series of plotted points and straight lines connecting the points or a smooth curve which has been "fitted" to the points by some method or other. This section will consider merely the information associated with the plotted points.

Figure 17 gives an example of such an observed relationship. (Data from records of shelf life tests on die-cast metals and alloys, former Subcommittee

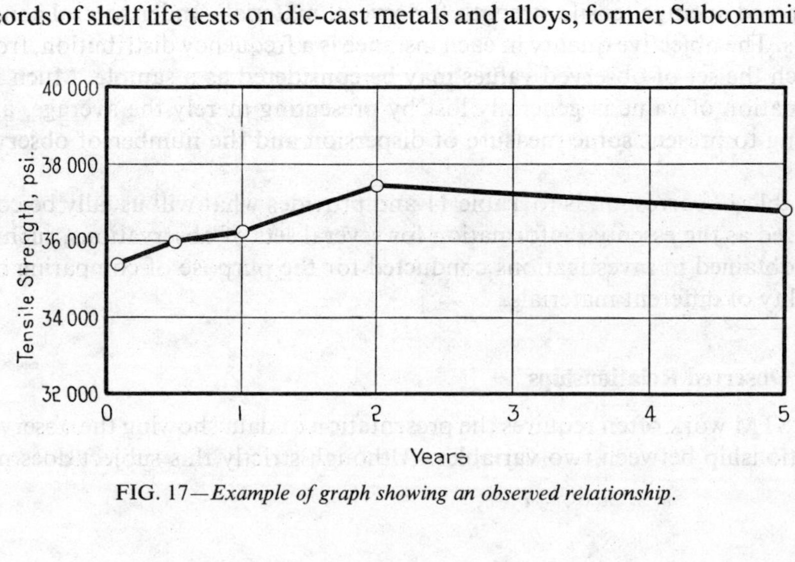

FIG. 17—*Example of graph showing an observed relationship.*

15 of ASTM Committee B-2 on Non-Ferrous Metals and Alloys.) At each successive stage of an investigation to determine the effect of aging on several alloys, five specimens of each alloy were tested for tensile strength by each of several laboratories. The curve shows the results obtained by one laboratory for one of these alloys. Each of the plotted points is the average of five observed values of tensile strength and thus attempts to summarize an observed frequency distribution. Figure 18 has been drawn to show pic-

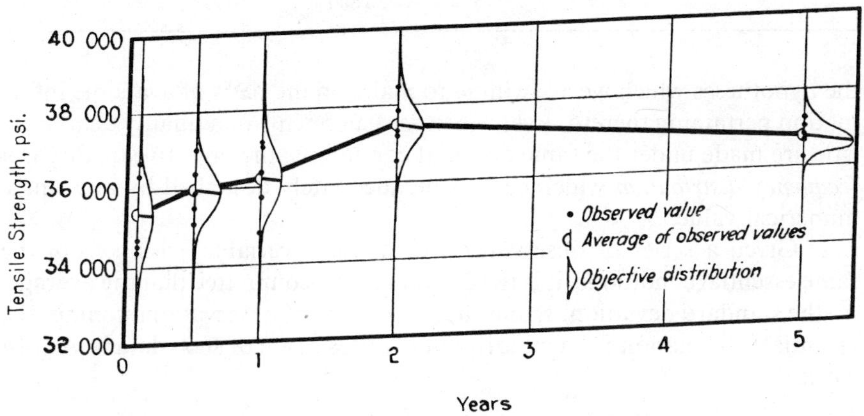

FIG. 18—*Showing pictorially what lies back of the plotted points in Fig. 17. Each plotted point in Fig. 17 is the average of a sample from a universe of possible observations.*

torially what is behind the scenes. The five observations made at each stage of the life history of the alloy constitute a sample from a universe of possible values of tensile strength—an objective frequency distribution whose spread is dependent on the inherent variability of the tensile strength of the alloy and on the error of testing. The dots represent the observed values of tensile strength and the bell-shaped curves, the objective distributions.

In such instances, the essential information contained in the data may be made available by supplementing the graph by a tabulation of the averages, the standard deviations, and the number of observations for the plotted points in the manner shown in Table 13.

41. Summary: Essential Information

The material given in Sections 37 to 40, inclusive, may be summarized as follows.

1. What constitutes the *essential information* in any particular instance depends on the nature of the questions to be answered, and on the nature of

TABLE 13—*Summary of essential information for Fig. 18.*

Time of Test	Number of Specimens	Tensile Strength, psi: Average, $\overline{X}$	Tensile Strength, psi: Standard Deviation, s
Initial	5	35 400	950
6 months	5	35 980	668
1 year	5	36 220	869
2 years	5	37 460	655
5 years	5	36 800	319

the hypotheses which we are willing to make on the basis of available information pertaining thereto. Even when measurements of a quality characteristic are made under the same essential conditions, the objective quality is a *frequency distribution* which cannot be adequately described by any single numerical value.

2. Given a series of observations of a single variable arising from the same essential conditions, it is the opinion of the committee that the average, $\overline{X}$, the standard deviation, s, and the number, n, of observations contain the essential information for a majority of the uses made of such data in ASTM work.

NOTE—If the observations are not obtained under the same essential conditions, analysis and and presentation by the control chart method, in which *order* (see Part 3 of this Manual) is taken into account by rational subgrouping of observations, commonly provides important additional information.

PRESENTATION OF RELEVANT INFORMATION

42. Introduction

Empirical knowledge is not contained in the observed data alone, rather it arises from interpretation—an act of thought. (For an important discussion on the significance of prior information and hypothesis in the interpretation of data, see Ref *10*; a treatise on the philosophy of probable inference which is of basic importance in the interpretation of any and all data is presented in Ref *11*.) Interpretation consists in testing hypotheses based on prior knowledge. Data constitute but a part of the information used in interpretation—the judgments that are made depend as well on pertinent collateral information, much of which may be of a qualitative rather than of a quantitative nature.

If the data are to furnish a basis for most valid prediction, they must be obtained under controlled conditions and must be free from constant errors of measurement. Mere presentation does not alter the goodness or badness of data. The usefulness of good data may, however, be enhanced by the manner in which they are presented.

43. Relevant Information

Presented data should be accompanied by available relevant information, particularly information on *precisely* the field within which the measurements are supposed to hold and the conditions under which they were made, and evidence that the data are good. Among the specific things that may be presented with ASTM data to assist others in interpreting them or to build up confidence in the interpretation made by an author are:

1. The kind, grade, and character of material or product tested.
2. The mode and conditions of production, if this has a bearing on the feature under inquiry.
3. The method of selecting the sample; steps taken to ensure its randomness or representativeness. (The manner in which the sample is taken has an important bearing on the interpretability of data and is discussed by Dodge [*12*].)
4. The specific method of test (if an ASTM or other standard test, so state; together with any modifications of procedure).
5. The specific conditions of test, particularly the regulation of factors that are known to have an influence on the feature under inquiry.
6. The precautions or steps taken to eliminate systematic or constant errors of observation.
7. The difficulties encountered and eliminated during the investigation.
8. Information regarding parallel but independent paths of approach to the end results.
9. Evidence that the data were obtained under controlled conditions; the results of statistical tests made to support belief in the constancy of conditions, in respect to the physical tests made or the material tested, or both. (Here, we mean constancy in the statistical sense, which encompasses the thought of stability of conditions from one time to another and from one place to another. This state of affairs is commonly referred to as "statistical control." Statistical criteria have been developed by means of which we may judge when controlled conditions exist. Their character and mode of application are given in Part 3 of this Manual; see also Ref *13*.)

Much of this information may be qualitative in character, some may even be vague, yet without it the interpretation of the data and the conclusions reached may be misleading or of little value to others.

44. Evidence of Control

One of the fundamental requirements of good data is that they should be obtained under controlled conditions. The interpretation of the observed results of an investigation depends on whether or not there is justification for believing that the conditions were controlled.

If the data are numerous and statistical tests for control are made, evidence of control may be presented by giving the results of these tests. (For examples, see Refs *14–17*.) Such quantitative evidence greatly strengthens inductive arguments. In any case, it is important to indicate clearly just what precautions were taken to control the essential conditions. Without tangible evidence of this character, the reader's degree of rational belief in the results presented will depend on his faith in the ability of the investigator to eliminate all causes of lack of constancy.

RECOMMENDATIONS

45. Recommendations for Presentation of Data

The following recommendations for presentation of data apply for the case where one has at hand a sample of n observations of a single variable obtained under the same essential conditions.

1. Present as a minimum, the average, the standard deviation, and the number of observations. *Always* state the number of observations taken.
2. If the number of observations is large, present also the value of the skewness, g_1, and the value of the kurtosis, g_2. An alternate procedure is to present a grouped frequency distribution.
3. If the data were not obtained under controlled conditions and it is desired to give information regarding the extreme observed effects of assignable causes, present the values of the maximum and minimum observations in addition to the average, the standard deviation, and the number of observations.
4. Present as much evidence as possible that the data were obtained under controlled conditions.
5. Present relevant information on precisely (*a*) the field within which the measurements are supposed to hold and (*b*) the conditions under which they were made.

SUPPLEMENTS

Supplement[5]

When the observations pertain to the "life" of a unit, it has been found that the Weibull distribution often gives a good approximation to the

[5] Added in 1975.

distribution of frequencies (see, for example, Ref *18*). The cumulative frequency of the Weibull distribution is given by the equation

$$F(X) = 1 - \exp\left[-\left(\frac{X - \alpha}{\gamma}\right)^{\beta} \right] \quad \alpha \le X < \infty \tag{15}$$

where β and γ are positive quantities and α can have any real value including 0. Since tables of the exponential function are readily available, $F(X)$ can be determined directly as soon as α, β, and γ are known.

The Weibull distribution can be positively skewed or negatively skewed. For values of β close to 3.26, it is almost Normal. When $\alpha = 0$ and $\beta = 1$, it becomes the negative exponential distribution discussed in Section 31.

To use Eq 15 as a graduating formula for a given set of data we need simply to estimate α, β, and γ. This can be done by using Weibull probability paper. A special variety of Weibull probability paper was presented by Nelson [*19*], and a copy of this is shown in Fig. 19. In his article, Nelson granted permission to reproduce this Weibull probability paper and a full page figure is presented in the journal for this purpose.

The Weibull probability paper illustrated in Fig. 19 is used as follows. Let X_i be the life of the i-th item after the items have been arranged in order from shortest to longest life. Subtract from each X_i a guessed value of α and plot $i/(n + 1)$ against $X_i - \alpha$. If the points follow a curve, vary α by trial and error until the points fall approximately on a straight line. The value of α that leads to this linear result is the graphical estimate of that parameter.

Next draw a line through the "estimation point" in the upper left-hand corner of the paper that is perpendicular to the straight line through the points on the chart. The intersection of this perpendicular with the scale marked $\hat{\beta}$ is an estimate of β. Finally the intercept of the straight line through the points with the vertical at the abscissa point $X - \alpha = 1$ yields an estimate of $\ln\gamma^{\beta}$ from which an estimate of γ may be obtained. Thus if $\beta = 1.7$ and the intercept $= 3.5$, we have $\ln(\gamma^{1.7}) = 3.5$ or $1.7 \ln\gamma = 3.5$ or $\ln\gamma = 2.06$ which yields $\gamma = e^{2.06} = 7.8$.

The most meaningful way to summarize a given sample of Weibull data is thus to present estimates of α, β and γ since approximations to the proportions of the data that lie within specified limits can be readily determined by substitution of these estimates in Eq 15.

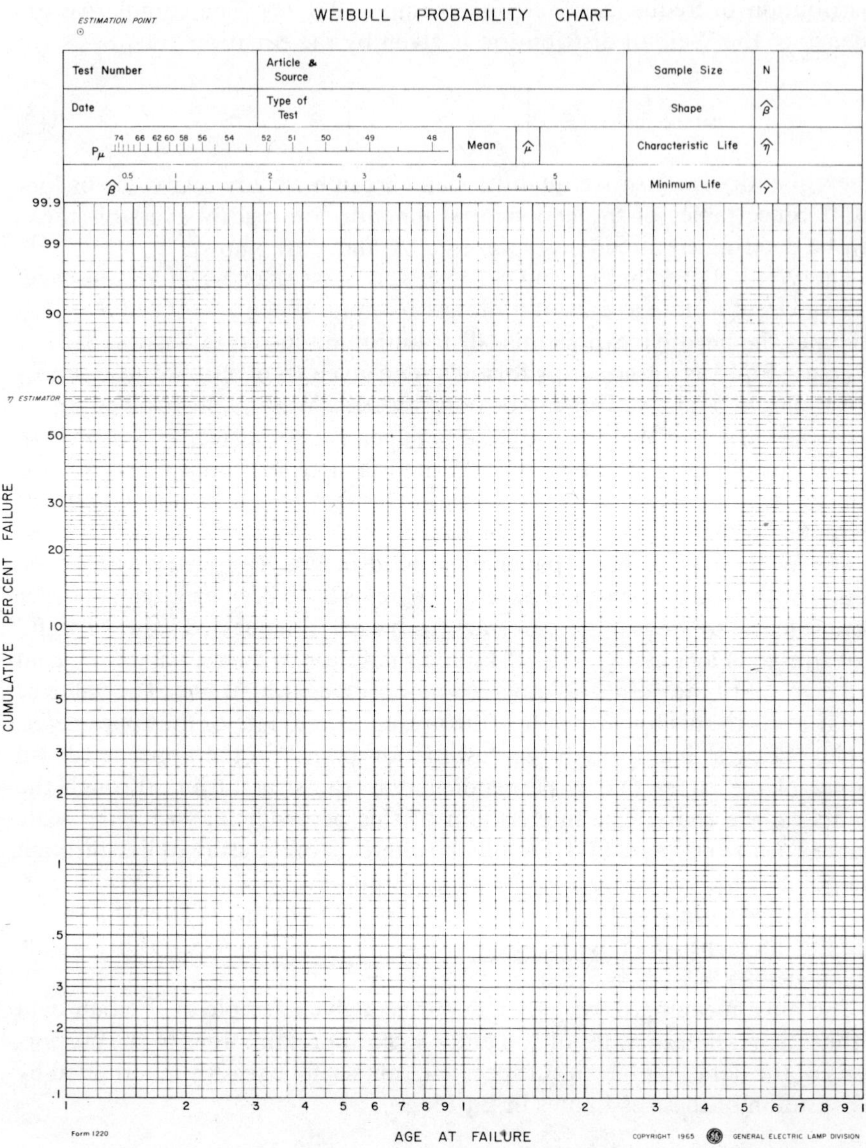

Reprinted by permission of the General Electric Lamp Division

FIG. 19—*Weibull probability chart.*

References

[*1*] Shewhart, W. A., *Economic Control of Quality of Manufactured Product,* Van Nostrand, New York, 1931.

[*2*] Yule, G. U. and Kendall, M. G. in *An Introduction to the Theory of Statistics*, 14th Ed., Charles Griffin and Company, Ltd., London, 1950, pp. 73–76.

[*3*] Hazen, Allen, *Transactions*, American Society of Civil Engineers, Vol. 77, 1914, p. 1539.

[*4*] Elderton, W. P. and Johnson, N. L., *Systems of Frequency Curves*, Cambridge University Press, Bentley House, London, 1969.

[*5*] Duncan, A. J., *Quality Control and Industrial Statistics*, 4th Ed., Chapter 6, Section 4, Richard D. Irwin, Inc., Homewood, Ill., 1974.

[*6*] Bowker, A. H. and Lieberman, G. J., *Engineering Statistics*, 2nd Ed., Section 8.12, Prentice-Hall, 1972.

[*7*] Hahn, G. J., *Journal of Quality Control Technology*, Vol. 2, No. 3, 1970, pp. 115–125.

[*8*] Tippett, L. H. C., *Biometrika*, Vol. 17, Dec. 1927, pp. 364–387.

[*9*] Hoel, P. G., *Introduction to Mathematical Statistics*, 4th Ed., Wiley, New York, 1971.

[*10*] Lewis, C. I., *Mind and the World Order*, Scribner, New York, 1929.

[*11*] Keynes, J. M., *A Treatise on Probability*, MacMillan, New York, 1921.

[*12*] Dodge, H. F., "Statistical Control in Sampling Inspection," presented at a round table discussion on "Acquisition of Good Data," held at the 1932 Annual Meeting of the American Society for Testing and Materials; published in *American Machinist*, 26 Oct. and 9 Nov. 1932.

[*13*] Pearson, E. S., *Journal of the Royal Statistical Society*, Vol. XCVI, Part 1, 1933, pp. 21–60.

[*14*] Passano, R. F., *Proceedings,* American Society for Testing Materials, Vol. 32, Part 2, 1932, p. 468.

[*14*] Skinker, M. F., *Proceedings*, American Society for Testing and Materials, Vol. 32, Part 2, 1932, p. 670.

[*16*] Passano, R. F. and Nagley, F. R., *Proceedings*, American Society for Testing and Materials, Vol. 33, Part 2, 1933, p. 387.

[*17*] Chancellor, W. C., *Proceedings*, American Society for Testing and Materials, Vol. 34, Part 2, 1934, p. 891.

[*18*] Hahn, G. J. and Shapiro, S. S., *Statistical Models in Engineering*, Wiley, New York, 1967, p. 108.

[*19*] Nelson, L., *Industrial Quality Control*, March 1967, pp. 452–453.

PART 2

Presenting Plus or Minus Limits of Uncertainty of an Observed Average

Presenting Plus or Minus Limits of Uncertainty of an Observed Average

Glossary of Symbols used in Part 2

a	Factor, given in Table 2 of Part 2, for computing confidence limits for $\overline{X}'$ associated with a desired value of probability, P, and a given number of observations, n
k	Deviation of a Normal variable
n	Number of observed values (observations)
P	Probability; used in Part 2 to designate the probability associated with confidence limits; relative frequency with which the averages, $\overline{X}'$, of sampled universes may be expected to be included within the confidence limits (for $\overline{X}'$) computed from samples
s	Sample standard deviation
$\hat{\sigma}$	Estimate of σ' based on several samples
X	Observed value of a measurable characteristic; specific observed values are designated X_1, X_2, X_3, etc.; also used to designate a measurable characteristic
$\overline{X}$	*Sample average* (arithmetic mean), the sum of the n observed values in a set divided by n
$\overline{X}'$, σ', etc.	True or objective value of the mean, standard deviation, etc., for the universe sampled (the prime (′) notation signifies the true or objective value as distinct from the observed value)

1. Purpose

Part 2 of the Manual discusses the problem of presenting plus or minus limits to indicate the uncertainty of the average of a number of observations obtained under the same essential conditions, and suggests a form of presentation for use in ASTM reports and publications where needed.

2. The Problem

An observed average, $\overline{X}$, is subject to the uncertainties that arise from sampling fluctuations, and tends to vary from the true average more widely if the number, n, of observations is small.

TABLE 1—*Breaking strength of 0.104-in. hard-drawn copper wire.*

Specimen	Breaking Strength, X, lb
1	578
2	572
3	570
4	568
5	572
6	570
7	570
8	572
9	576
10	584
$n = 10$	5732
Average, $\bar{X}$	573.2
Standard deviation, s	4.83

Having a set of n observed values of a variable X whose average (arithmetic mean) is $\bar{X}$, as in Table 1, it is often desired to present the results as

$$\text{average, } \bar{X} = 573.2 \pm 3.5 \text{ lb}$$

the plus or minus limits being established from the quantitative data alone with the implications that the objective average, $\bar{X}'$, of the universe sampled is included within such limits. How should such limits be computed, and what meaning may be attached to them?

NOTE—The objective average, $\bar{X}'$, is the value of $\bar{X}$ that would be approached as a statistical limit as more and more observations were obtained under the same essential conditions and their cumulative averages computed.

3. Theoretical Background

Mention should be made of the practice, now mostly out of date in scientific work, of recording such limits as

$$\bar{X} \pm 0.6745 \frac{s}{\sqrt{n}}$$

where

$\bar{X}$ = observed average,
s = observed standard deviation, and
n = number of observations,

and referring to the value 0.6745 $s/\sqrt{n}$ as the "probable error" of the observed average, $\bar{X}$, (the value of 0.6745 corresponding to the Normal law probability of 0.50). The term "probable error" and the probability value of 0.50 properly apply to the errors of sampling when sampling from a universe whose average, $\bar{X}'$, and whose standard deviation, σ', are *known* (these terms apply to limits $\bar{X}' \pm 0.6745\, \sigma'/\sqrt{n}$), but they do not apply in the inverse problem when merely sample values of $\bar{X}$ and s are given.

Investigation of this problem [*1–3*][1] has given a more satisfactory alternative (Section 4), a procedure which provides limits that have a definite operational meaning.

NOTE—While the method of Section 4 represents the best that can be done at present in interpreting a sample $\bar{X}$ and s when no other information regarding the variability of the universe is available, in general a much more satisfactory interpretation can be made if other information regarding the variability of the universe is at hand, such as a series of samples from the universe or similar universes for each of which a value of s or R is computed. If s or R displays statistical control as outlined in Part 3 of this Manual and a sufficient number of samples (preferably 20 or more) are available to obtain a reasonably precise estimate of σ', designated as $\hat{\sigma}$, the limits of uncertainty for a sample containing any number of observations, n, and arising from a universe whose true standard deviation can be presumed equal to $\hat{\sigma}$, can be computed from the following formula:

$$\bar{X} \pm k \frac{\hat{\sigma}}{\sqrt{n}}$$

where: k = 1.645, 1.960, and 2.576 for probabilities of P = 0.90, 0.95, and 0.99, respectively.

4. Computation of Limits

The following procedure applies to any long run *series* of problems for *each* of which the following conditions are met:

Given: A sample of n observations, $X_1, X_2, X_3, \ldots, X_n$, having an average = $\bar{X}$ and a standard deviation = s.

Conditions: (*a*) The universe sampled is homogeneous (statistically controlled) in respect to X, the variable measured. (*b*) The distribution of X for the universe sampled is approximately Normal. (*c*) The sample is a random sample.[2]

Procedure: Compute limits

$$\bar{X} \pm as$$

where the value of a is given in Table 2 for three values of P and for various values of n.

[1] The italic numbers in brackets refer to the list of references appended Part 2.

[2] If the universe sampled is *limited*, that is, made up of a limited number of separate units that may be measured in respect to the variable, X, and if interest centers on the $\bar{X}'$ of this limited universe, then this procedure assumes that the number of units, n, in the sample is relatively small compared with the number of units, N, in the universe, say n is less than about 5 percent of N. However, correction for relative size of sample can be made by multiplying s by the factor $\sqrt{1 - (n/N)}$. On the other hand, if interest centers on the $\bar{X}'$ of the underlying process or source of the limited universe, then this correction factor is not used.

TABLE 2—*Factors for calculating 90, 95, and 99 percent confidence limits for averages.*[a]

Number of Observations in Sample, n	Confidence Limits,* $\overline{X} \pm as$		
	90% Confidence Limits ($P = 0.90$) Value of a	95% Confidence Limits ($P = 0.95$) Value of a	99% Confidence Limits ($P = 0.99$) Value of a
4	1.177	1.591	2.921
5	0.953	1.241	2.059
6	0.823	1.050	1.646
7	0.734	0.925	1.401
8	0.670	0.836	1.237
9	0.620	0.769	1.118
10	0.580	0.715	1.028
11	0.546	0.672	0.955
12	0.518	0.635	0.897
13	0.494	0.604	0.847
14	0.473	0.577	0.805
15	0.455	0.554	0.769
16	0.438	0.533	0.737
17	0.423	0.514	0.708
18	0.410	0.497	0.683
19	0.398	0.482	0.660
20	0.387	0.468	0.640
21	0.376	0.455	0.621
22	0.367	0.443	0.604
23	0.358	0.432	0.588
24	0.350	0.422	0.573
25	0.342	0.413	0.559
n greater than 25	$a = \frac{1.645}{\sqrt{n}}$ approximately	$a = \frac{1.960}{\sqrt{n}}$ approximately	$a = \frac{2.576}{\sqrt{n}}$ approximately

*Limits which may be expected to include $\overline{X}'$ (9 times in 10.95 times in 100, or 99 times in 100) in a series of problems, each involving a single sample of n observations.

Values of a computed from Fisher, R. A., "Table of t," *Statistical Methods for Research Workers*, Table 4, based on student's distribution of z.

[a] Recomputed in 1975. The a of this Table 2 equals Fisher's t for $n - 1$ degrees of freedom divided by $\sqrt{n}$. See also Fig. 1.

Meaning: If the values of a given in Table 2 for $P = 0.95$ are used in a series of such problems, then, in the long run, we may expect 95 percent of the ranges bounded by the limits so computed, to include the objective averages, $\overline{X}'$, of the universes sampled. If in each instance, we were to assert that $\overline{X}'$ is included within the limits computed, we should expect to be correct 95 times in 100 and in error 5 times in 100; that is, the statement "$\overline{X}'$ is included within the range so computed" has a probability of 0.95 of being

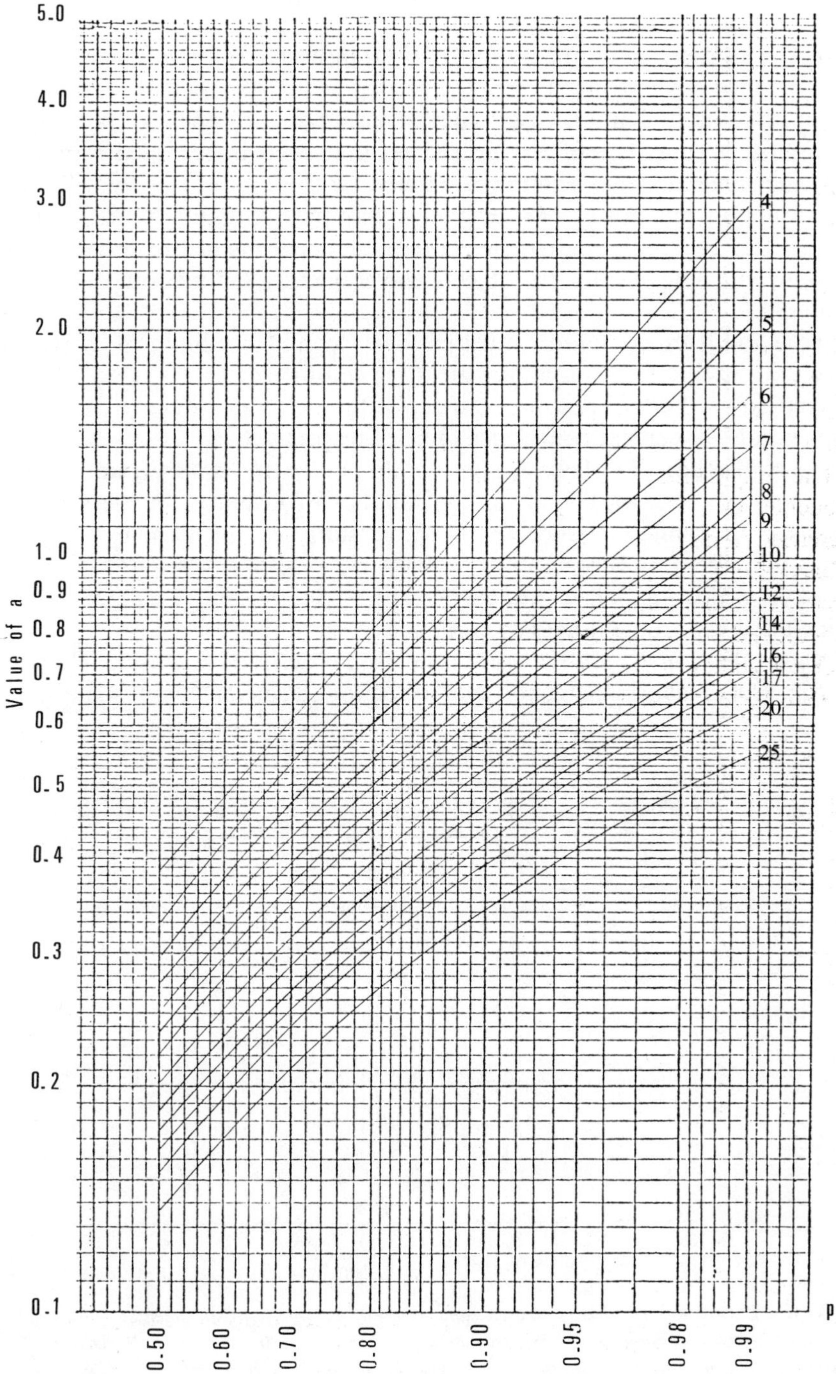

FIG. 1—*Curves giving factors for calculating 50 to 99 percent confidence limits for averages (see also Table 2). Redrawn in 1975 for new values of* a. *Error in reading* a *not likely to be* >0.01.

correct. But, there would be no operational meaning in the following statement made in any one instance: "The probability is 0.95 that $\bar{X}'$ falls within the limits computed in this case" since $\bar{X}'$ either does or does not fall within the limits. It should also be emphasized that even in repeated sampling from the *same* universe, the range defined by the limits $\bar{X} \pm as$ will vary in width and position from sample to sample, particularly with small samples (see Fig. 2). It is this series of ranges fluctuating in size and position which will include, ideally, the objective average, $\bar{X}'$, 95 times out of 100 for $P = 0.95$.

These limits are commonly referred to as "confidence limits" (see Ref *3*, pp. 65–71, and Ref *4*, p. 227); for the three columns of Table 2 they may be referred to as the "90 percent confidence limits," "95 percent confidence limits" and "99 percent confidence limits," respectively.

The magnitude $P = 0.95$ applies to the *series of instances*, and is approached as a statistical limit as the number of instances in the series is increased indefinitely; hence it signifies "statistical probability." If the values of a given in Table 2 for $P = 0.99$ are used in a series of instances, we may, in like manner, expect 99 percent of the universe averages, $\bar{X}'$, to be included within the ranges so computed.

Other values of P could, of course, be used if desired—the use of chances of 95 in 100, or 99 in 100 are, however, often found convenient in engineering

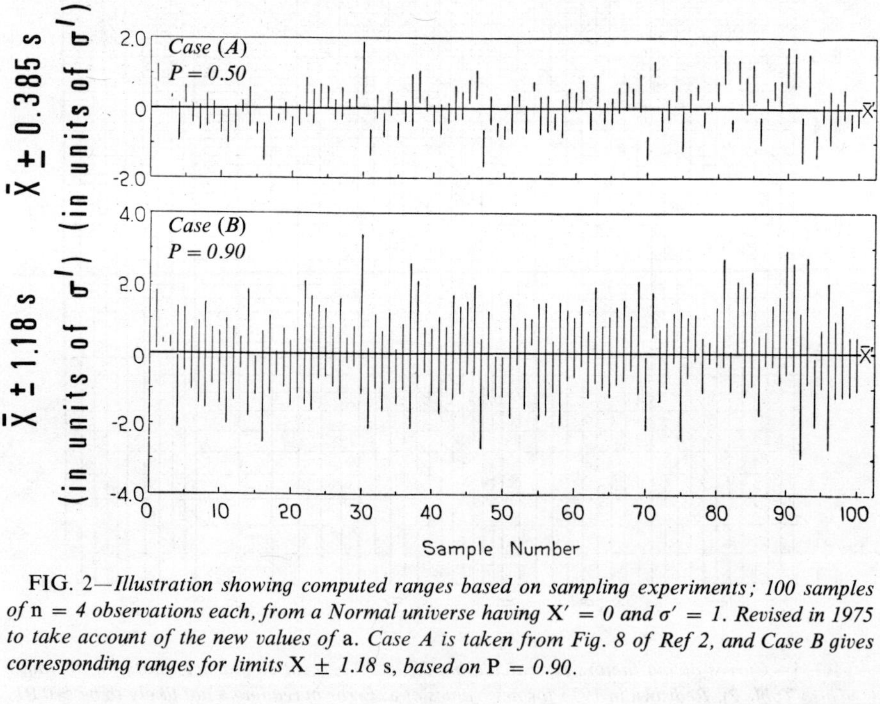

FIG. 2—*Illustration showing computed ranges based on sampling experiments; 100 samples of* n = *4 observations each, from a Normal universe having* X′ = *0 and* σ′ = *1. Revised in 1975 to take account of the new values of* a. *Case A is taken from Fig. 8 of Ref 2, and Case B gives corresponding ranges for limits* X ± *1.18* s, *based on* P = *0.90.*

presentations. Approximate values of a for other values of P may be read from the curves in Fig. 1, for samples of $n = 25$ or less.

For larger samples (n greater than 25), the constants, 1.645, 1.960, and 2.576, in the expressions

$$a = \frac{1.645}{\sqrt{n}}, \quad a = \frac{1.960}{\sqrt{n}}, \quad \text{and} \quad a = \frac{2.576}{\sqrt{n}}$$

at the foot of Table 2 are obtained directly from Normal law integral tables for probability values of 0.90, 0.95, and 0.99. To find the value of this constant for any other value of P, consult any standard text on statistical methods or read the value approximately on the "k" scale of Fig. 15 of Part 1 of this Manual.

5. Experimental Illustration

Figure 2 gives two diagrams illustrating the results of sampling experiments for samples of $n = 4$ observations each drawn from a Normal universe, for values of Case A, $P = 0.50$, and Case B, $P = 9.90$. For Case A, the ranges for 51 out of 100 samples included $\bar{X}'$ and for Case B, 90 out of 100 included $\bar{X}'$. If, in each instance (that is, for each sample) we had concluded that the objective $\bar{X}'$ is included within the limits shown for Case A, we would have been correct 51 times and in error 49 times, which is a reasonable variation from the expectancy of being correct 50 percent of the time.

In this experiment all samples were taken from the same universe. However, it will be obvious that the same reasoning would apply to a series of samples each taken from a different universe provided the conditions of Section 4 are met.

6. Presentation of Data

In presentation of data, if it is desired to give limits of this kind, it is quite important that the significance of the limits be clearly indicated. The three values $P = 0.90$, $P = 0.95$, and $P = 0.99$ given in Table 2 (chances of 9 in 10, 95 in 100, and 99 in 100) are arbitrary choices that may be found convenient in practice.

Example

Given a sample of 10 observations of breaking strength of hard-drawn copper wire as in Table 1, for which

$$\bar{X} = 573.2 \text{ lb}$$
$$s = 4.83 \text{ lb}$$

Using this sample to define limits of uncertainty based on $P = 0.95$ (Table 2), we have

$$\begin{aligned}\bar{X} \pm 0.715s &= 573.2 \pm 3.5 \\ &= 569.7 \text{ and } 576.7\end{aligned}$$

Two pieces of information are needed to supplement this numerical result: (*a*) the fact that 95 in 100 limits were used, and (*b*) that this result is based solely on the evidence contained in 10 observations.

Hence, in the presentation of such limits, it is desirable to give the results in some such way as the following

$$573.2 \pm 3.5 \text{ lb } (P = 0.95, n = 10)$$

The essential information contained in the data is, of course, covered by presenting $\bar{X}$, s, and n (see Part 1 of this Manual) and the limits under discussion could be derived directly therefrom. If it is desired to present such limits in addition to $\bar{X}$, s, and n, the tabular arrangement given as follows is suggested.

Number of Tests, n	Average, $\bar{X}$	Limits for $\bar{X}'$ (95 percent Confidence Limits)	Standard Deviation, s
10	573.2	573.2 ± 3.5	4.83

A satisfactory alternative is to give the ± value in the column designated "Average," and to add a note giving the significance of this entry, as follows.

Number of Tests, n	Average,[a] $\bar{X}$	Standard Deviation, s
10	573.2(±3.5)	4.83

[a]The ± entry indicates 95 percent confidence limits for $\bar{X}'$.

6a. One-Sided Limits

Sometimes we are interested in limits of uncertainty in one direction only. In this case we would present $\bar{X} + as$ or $\bar{X} - as$ (not both), indicating that this represents a one-sided confidence limit to contain the universe mean. The a to use in this one-sided case and the associated confidence coefficient would be obtained from Table 2 or Fig. 1 as follows.

For a confidence coefficient of 0.95 use the a listed in Table 2 under $P = 0.90$.

For a confidence coefficient of 0.975 use the a listed in Table 2 under $P = 0.95$.

For a confidence coefficient of 0.995 use the a listed in Table 2 under $P = 0.99$.

In general for a confidence coefficient of P_1 use the a derived from Fig. 1 for $P = 1 - 2(1 - P_1)$. For example, with $n = 10$, $\overline{X} = 573.2$, and $s = 4.83$ the one-sided upper 0.95 confidence limit would be $573.2 + 0.58(4.83) = 573.2 + 2.8 = 576.0$.

7. Number of Places to be Retained in Computation and Presentation

The following working rule is recommended in carrying out computations incident to determining averages, standard deviations, and "limits for averages" of the kind here considered, for a sample of n observed values of a variable quantity:

In all operations on the sample of n observed values, such as adding, subtracting, multiplying, dividing, squaring, extracting square root, etc., retain the equivalent of two more places of figures than in the single observed values. For example, if observed values are read or determined to the nearest 1 lb, carry numbers to the nearest 0.01 lb in the computations; if observed values are read or determined to the nearest 10 lb, carry numbers to the nearest 0.1 lb in the computations, etc.

Deleting places of figures should be done after computations are completed, in order to keep the final results substantially free from computation errors. In deleting places of figures the actual rounding-off procedure should be carried out as follows.[3]

1. When the figure next beyond the last figure or place to be retained is less than 5, the figure in the last place retained should be kept unchanged.
2. When the figure next beyond the last figure or place to be retained is more than 5, the figure in the last place retained should be increased by 1.
3. When the figure next beyond the last figure or place to be retained is 5, and (*a*) there are no figures, or only zeros, beyond this 5, if the figure in the last place to be retained is odd, it should be increased by 1; if even, it should be kept unchanged; but (*b*) if the 5 next beyond the figure in the last place to be retained is followed by any figures other than zero, the figure in the last place retained should be increased by 1, whether odd or even.

For example, if in the following numbers, the places of figures in parenthesis are to be rejected

39 4(49) becomes 39 400
39 4(50) becomes 39 400,
39 4(51) becomes 39 500, and
39 5(50) becomes 39 600.

[3] This rounding-off procedure agrees with that adopted in the American Standard Rules for Rounding-Off Numerical Values (ANSI Project: Z25.1-1940 (R1961)).

The number of places of figures to be retained in presentation depends on what use is to be made of the results and on the sampling variation present. No general rule, therefore, can safely be laid down. The following working rule has, however, been found generally satisfactory by the committee in presenting the results of testing in technical investigations and development work.

(*a*) For averages

When the Single Values are Obtained to the Nearest	and the	Number of Observed Values is	
0.1, 1, 10, etc., units		2 to 20	21 to 200
0.2, 2, 20, etc., units	less than 4	4 to 40	41 to 400
0.5, 5, 50, etc., units	less than 10	10 to 100	101 to 1000
Retain the following number of places of figures in the average	Same number of places as in single values	1 more place than in single values	2 more places than in single values

(*b*) For standard deviations, retain three places of figures.

(*c*) If "limits for averages" of the kind here considered are presented, retain the same places of figures as are retained for the average.

For example, if $n = 10$, and if observed values were obtained to the nearest 1 lb, present averages and "limits for averages" to the nearest 0.1 lb, and present the standard deviation to three places of figures. This is illustrated in the tabular presentation in Section 6.

Rule (*a*) will result generally in one and conceivably in two doubtful places of figures in the average, that is, places which may have been affected by the rounding-off (or observation) of the n individual values to the nearest number of units stated in the first column of the table. Referring to the tabular arrangement in Section 6, the third place of figures in the average, $\overline{X} = 573.2$, corresponding to the first place of figures in the $\pm$ 3.5 value is doubtful in this sense. One might conclude that it would be suitable to present the average to the nearest pound, thus

$$573 \pm 3 \text{ lb } (P = 0.95, n = 10)$$

This might be satisfactory for some purposes. However, the effect of such rounding off to the first place of figures of the plus or minus value may be quite pronounced if the first digit of the plus or minus value is small, as indicated in the following table.

	Not Rounded				Rounded		
	Limits		Difference		Limits		Difference
573.5 ± 1.4	572.1	574.9	2.8	574 ± 1	573	575	2
573.5 ± 1.5	572.0	575.0	3.0	574 ± 2	572	576	4

If further use were to be made of these data—accumulating additional observations to be combined with these, gathering other data to be compared with these, etc.—then the effect of such rounding off of $\bar{X}$ in a presentation might seriously interfere with proper subsequent use of the information.

The number of places of figures to be retained or to be used as a basis for action in specific cases (such as in reports covering the acceptance and rejection of material) cannot readily be made subject to any general rule. It is, therefore, recommended that in such cases the number of places be settled by definite agreements between the individuals or parties involved. The ASTM Recommended Practices for Indicating Which Places are to be Considered Significant in Specified Limiting Values (E 29-67(1973)) give specific rules which are applicable when reference is made to these recommended practices.

8. General Comments on the Use of Confidence Limits

In making use of limits of uncertainty of the type covered in this part, the engineer should keep in mind: (1) the restrictions as to (*a*) controlled conditions, (*b*) approximate Normality of universe, (*c*) randomness of sample; and (2) the fact that the variability under consideration relates to fluctuations around the level of measurement values, whatever that may be, quite regardless of whether or not the objective average, $\bar{X}'$, of the measurement values is widely displaced from the true value, $\bar{X}'_T$, of the thing measured, as a result the systematic or *constant* errors present throughout the measurements.

For example, breaking strength values might center around a value of 575.0 lb (the objective level $\bar{X}'$ of the measurement values) with a scatter of individual observations represented by the dotted distribution curve of Fig. 3, whereas the true average $\bar{X}'_T$ for the batch of wire under test is actually 610.0 lb, the difference between 575.0 and 610.0 representing a constant or systematic error present in *all* the observations as a result, say, of an incorrect adjustment of the testing machine.

The limits thus have meaning for series of like measurements, made under like conditions, *including* the same constant errors if any be present.

In the practical use of these limits, the engineer may not have assurance that conditions (*a*), (*b*), and (*c*) given in Section 4 are met, hence it is not advisable to lay great emphasis on the exact magnitudes of the probabilities

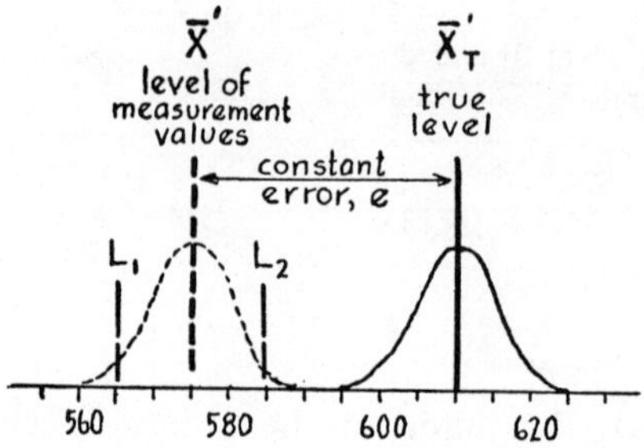

FIG. 3—*Showing how plus or minuts limits* (L_1 *and* L_2) *are unrelated to a systematic or constant error.*

given in Table 2, but rather to consider them as orders of magnitude to be used as general guides.

Supplement A[4]

Presenting Plus or Minus Limits of Uncertainty for σ', Normal Universe[5]

When observations $X_1, \ldots, X_n$ are made under controlled conditions and there is reason to believe the distribution of X is Normal, confidence limits for the standard deviation of the universe with confidence coefficient, P, will be given by

$$\begin{aligned} \text{lower confidence limit for } \sigma' &= s\sqrt{(n-1)/\chi^2_{(1-P)/2}} \\ \text{upper confidence limit for } \sigma' &= s\sqrt{(n-1)/\chi^2_{(1+P)/2}} \end{aligned} \tag{1}$$

where the quantities $\chi^2_{(1-P)/2}$ and $\chi^2_{(1+P)/2}$ will be found in most statistics text books.

To facilitate computation, Table 3 gives values of

$$b_L = \sqrt{(n-1)/\chi^2_{(1-P)/2}} \quad \text{and} \quad b_U = \sqrt{(n-1)/\chi^2_{(1+P)/2}} \tag{2}$$

for $P = 0.90$, 0.95, and 0.99. Thus we have, for a Normal universe

$$\text{lower confidence limit for } \sigma' = b_L s$$

and

$$\text{upper confidence limit for } \sigma' = b_U s \tag{3}$$

Example—Table 5 of Part 1 gives the standard deviation of a sample of 10 observations of breaking strength of copper wire as $s = 4.83$ lb. If we

[4] Added in 1975.

[5] The analysis is strictly valid only for an unlimited universe such as represented by a manufacturing or measurement process. When the universe sampled is relatively small compared with the sample size n, the reader is advised to consult a statistician.

TABLE 3—b-*factors for calculating confidence limits for* σ', *Normal universe.*[a]

Number of Observations in Sample, n	90% Confidence Limits b_L	b_U	95% Confidence Limits b_L	b_U	99% Confidence Limits b_L	b_U
2	0.510	16.0	0.446	31.9	0.356	159.5
3	0.578	4.41	0.521	6.29	0.434	14.1
4	0.619	2.92	0.566	3.73	0.484	6.47
5	0.649	2.37	0.600	2.87	0.518	4.40
6	0.671	2.09	0.625	2.45	0.547	3.48
7	0.690	1.91	0.645	2.20	0.569	2.98
8	0.705	1.80	0.661	2.04	0.587	2.66
9	0.718	1.71	0.676	1.92	0.603	2.44
10	0.730	1.64	0.688	1.83	0.618	2.28
11	0.739	1.59	0.698	1.75	0.630	2.15
12	0.747	1.55	0.709	1.70	0.641	2.06
13	0.756	1.51	0.718	1.65	0.651	1.98
14	0.762	1.49	0.725	1.61	0.660	1.91
15	0.769	1.46	0.732	1.58	0.669	1.85
16	0.775	1.44	0.739	1.55	0.676	1.81
17	0.780	1.42	0.745	1.52	0.683	1.76
18	0.785	1.40	0.750	1.50	0.690	1.73
19	0.789	1.38	0.756	1.48	0.696	1.70
20	0.794	1.37	0.760	1.46	0.702	1.67
21	0.798	1.35	0.765	1.44	0.707	1.64
22	0.801	1.35	0.769	1.43	0.712	1.62
23	0.806	1.34	0.773	1.41	0.717	1.60
24	0.808	1.33	0.777	1.40	0.721	1.58
25	0.812	1.32	0.780	1.39	0.725	1.56
26	0.814	1.31	0.785	1.38	0.730	1.54
27	0.818	1.30	0.788	1.37	0.734	1.52
28	0.821	1.29	0.791	1.36	0.738	1.51
29	0.823	1.29	0.793	1.35	0.741	1.50
30	0.825	1.28	0.797	1.35	0.745	1.49
31	0.828	1.27	0.799	1.34	0.747	1.47
For larger n and	$1/(1+1.645/\sqrt{2n})$	$1/(1-1.645/\sqrt{2n})$	$1/(1+1.960/\sqrt{2n})$	$1/(1-1.960/\sqrt{2n})$	$1/(1+2.576/\sqrt{2n})$	$1/(1-2.576/\sqrt{2n})$

[a] Confidence limits for $\sigma' = b_L s$ and $b_U s$.

assume that the breaking strength has a Normal distribution, which may actually be somewhat questionable, we have as 0.95 confidence limits for the universe standard deviation

$$\text{lower 0.95 confidence limit for } \sigma' = 0.688(4.83) = 3.32 \text{ lb}$$

$$\text{upper 0.95 confidence limit for } \sigma' = 1.83(4.83) = 8.83 \text{ lb}$$

If we wish a one-sided confidence limit on the low side with confidence coefficient, P, we take

$$\text{lower one-sided confidence limit for } \sigma' = s\sqrt{(n-1)/\chi^2_{(1-P)}}$$

For a one-sided confidence limit on the high side with confidence coefficient P, we take

$$\text{upper one-sided confidence limit for } \sigma' = s\sqrt{(n-1)/\chi^2_P}$$

Thus for $P = 0.95$, 0.975, and 0.995 we use the b_L or b_U factor from Table 3 in the columns headed 0.90, 0.95, and 0.99, respectively. For example, a 0.95 upper one-sided confidence limit for σ' based on a sample of 10 items for which $s = 4.83$ will be

$$\text{upper 0.95 one-sided confidence limit} = b_{U(0.90)}s = 1.64(4.83) = 7.92$$

A lower 0.95 one-sided confidence limit would be

$$\text{lower 0.95 one-sided confidence limit} = b_{L(0.90)}s = 0.730(4.83) = 3.53$$

Supplement B[6]

Presenting Plus or Minus Limits of Uncertainty for p′

When there is a fraction p of a given category, for example, the fraction defective, in n observations obtained under controlled conditions, 95 percent confidence limits for the universe fraction p' (see footnote 5) may be found from Fig. 41 of *Biometrika Tables for Statisticians*, Vol. 1. A reproduction of this for selected sample sizes is shown in Fig. 4. To use the Biometrika chart, the sample fraction is entered on the abscissa and the upper and lower 0.95 confidence limits are read on the vertical scale for various values of n. Approximate limits for values of n not shown on the Biometrika chart may be obtained by graphical interpolation. The *Biometrika Tables for Statisticians* also give a chart for 0.99 confidence limits.

For numerical values for 0.80, 0.95, and 0.99 confidence limits for p', the reader is referred to Tables 5 to 8 of Ref. 5.

In general, for an np at least ≥ 5 and preferably $p \geq 0.10$, the following formulas can be applied

$$\text{approximate 0.90 confidence limits} = p \pm 1.645\sqrt{p(1-p)/n}$$

$$\text{approximate 0.95 confidence limits} = p \pm 1.960\sqrt{p(1-p)/n} \qquad (4)$$

$$\text{approximate 0.99 confidence limits} = p \pm 2.576\sqrt{p(1-p)/n}.$$

Example—Refer to the data of Table 1(*a*) of Part 1 and Fig. 4 of Part 1 and suppose that the lower specification limit on transverse strength is

[6] Added in 1975.

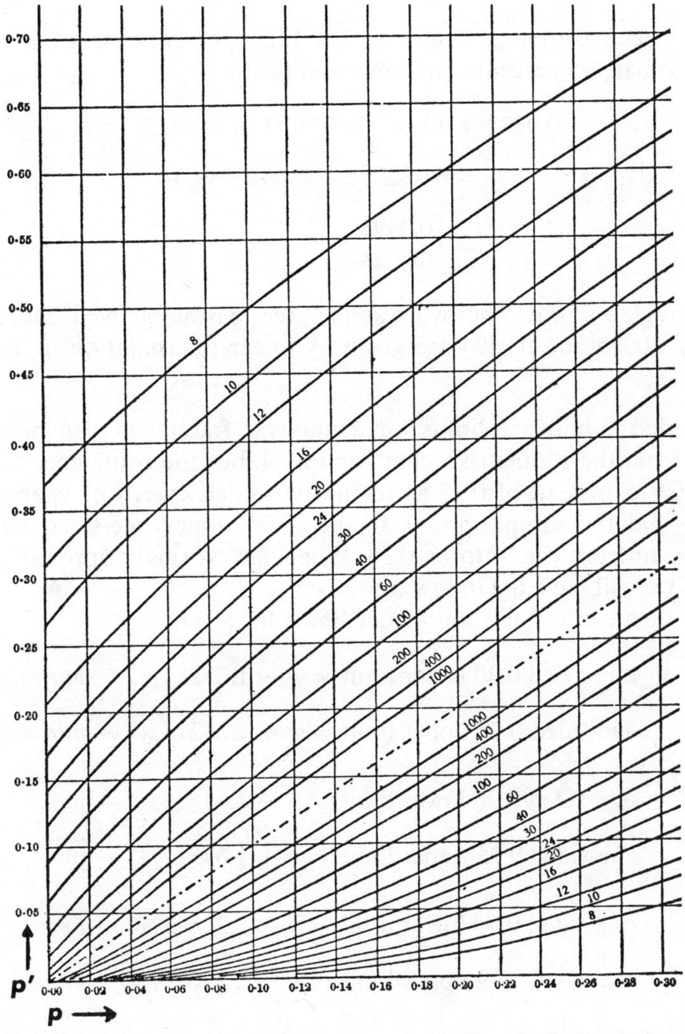

Reproducted by permission of the Biometrika Trust

FIG. 4—*Chart providing confidence limits for* p′ *in binomial sampling, given a sample fraction. Confidence coefficient = 0.95. The numbers printed along the curves indicate the sample size* n. *If for a given value of the abscissa,* p_A *and* p_B *are the ordinates read from (or interpolated between) the appropriate lower and upper curves, then* $\Pr\{p_A \leq p' \leq p_B\} \geq 0.95$.

675 psi and there is no upper specification limit. Then the sample percentage of bricks nonconforming (the sample "fraction defective," *p*) is seen to be 6/270 = 2.22 percent. Rough 0.95 confidence limits for the universe

fraction defective are read from Fig. 4 as 1 to 5 percent. Using Eq 4, we have approximate 95 percent confidence limits are

$$0.0222 \pm 1.960\sqrt{0.0222(1 - 0.0222)/270}$$
$$= 0.0222 \pm 1.960(0.00897)$$
$$= \begin{cases} 0.0498 \\ 0.0046 \end{cases}$$

Even though $p < 0.10$, the two results agree reasonably well. More precise limits of 0.0085 and 0.0494 are given by linear interpolation in Table 5 of Ref 5, p. 26.

One-sided confidence limits for a universe fraction p' can be obtained directly from the Biometrika chart or Fig. 4, but the confidence coefficient will be 0.975 instead of 0.95 as in the two-sided case. For example, with $n = 200$ and the sample $p = 0.10$, the 0.975 upper one-sided confidence limit is read from Fig. 4 to be 0.15. When the Normal approximation can be used, we will have the following.
approximate one-sided confidence limits for p' will be

$$\text{one-sided 0.90 lower limit} = p - 1.282\sqrt{p(1 - p)/n}$$
$$\text{one-sided 0.90 upper limit} = p + 1.282\sqrt{p(1 - p)/n}$$

$$\text{one-sided 0.95 lower limit} = p - 1.645\sqrt{p(1 - p)/n}$$
$$\text{one-sided 0.95 upper limit} = p + 1.645\sqrt{p(1 - p)/n}$$

$$\text{one-sided 0.99 lower limit} = p - 2.326\sqrt{p(1 - p)/n}$$
$$\text{one-sided 0.99 upper limit} = p + 2.326\sqrt{p(1 - p)/n}$$

References

[1] Shewhart, W. A., "Probability as a Basis for Action," presented at the joint meeting of the American Mathematical Society and Section K of the A. A. A. S., 27 Dec. 1932.

[2] Shewhart, W. A., "Statistical Method from the Viewpoint of Quality Control," W. E. Deming, Ed., The Graduate School, Department of Agriculture, Washington, D. C., 1939.

[3] Pearson, E. S., "The Application of Statistical Methods to Industrial Standardization and Quality Control," B. S. 600-1935, British Standards Institution, London, Nov. 1935.

[4] Hoel, P. G., *Introduction to Mathematical Statistics,* 4th Ed., Wiley, New York, 1971.

[5] Mainland, D., Herrera, L., and Sutcliffe, M. I., *Tables for Use with Binomial Samples,* Department of Medical Statistics, New York University, College of Medicine, 1956.

PART 3

Control Chart Method of Analysis and Presentation of Data

Control Chart Method of Analysis and Presentation of Data

Glossary of Terms and Symbols used in Part 3

In general, the terms and symbols used in Part 3 have the same meanings as in preceding parts of the Manual. In a few cases, which are indicated in the following glossary, a more specific meaning is attached to them for the convenience of a portion or all of Part 3. Mathematical definitions and derivations are given in Supplement A.

Glossary of Terms

assignable cause, *n.*—an identifiable factor that contributes to variation in quality.

lot, *n.*—a definite quantity of some commodity manufactured under similar conditions of production.

sample, *n.*—a group of units, or portion of material, taken from a larger collection of units or quantity of material, which serves to provide information that can be used as a basis for action on the larger quantity or on the production process. The term is also used in the sense of a sample of observations.

subgroup, *n.*—one of a series of groups of observations obtained by subdividing a larger group of observations; alternatively, the data obtained from one of a series of samples taken from a series of lots or from a process. One of the essential features of the control chart method is to break up the inspection data into *rational subgroups*, that is, to classify the observed values into subgroups, *within* which variations may for engineering reasons be considered to be due to nonassignable chance causes only, but *between* which there may be differences due to assignable causes whose presence is considered possible.

unit, *n.*—one of a number of similar articles, parts, specimens, lengths, areas, etc., of a material or product.

Glossary of Symbols

Symbol	General	In Part 3, Control Charts
c		*number of defects;* more specifically the number of defects in a sample (subgroup)

Symbol	General	In Part 3, Control Charts
c_4		a factor that is a function of and expresses the ratio between the expected value of $\bar{s}$ for a large numbur of samples of n observed values each and the σ' of the universe sampled. (Values of $c_4 = E(\bar{s})/\sigma'$ are given in Tables 2 and 3, and in Table 28 in Supplement A based on a Normal distribution.)
d_2		a factor that is a function of n and expresses the ratio between the expected value of $\bar{R}$ for a large number of samples of n observed values each and the σ' of the universe sampled. (Values of $d_2 = E(\bar{R})/\sigma'$ are given in Tables 2 and 3 and in Table 28 in Supplement A, based on a Normal distribution.)
k		number of subgroups or samples under consideration
n	*number* of observed values (observations)	subgroup or sample size, that is, the number of units or observed values in a sample or subgroup
p	*relative frequency or proportion*, the ratio of the number of occurrences to the total possible number of occurrences	*fraction defective*, the ratio of the number of defective units (articles, parts, specimens, etc.) to the total number of units under consideration; more specifically, the fraction defective of a sample (subgroup)
pn	number of occurrences	*number of defectives* (defective units); more specifically, the *number of defectives in* a sample of n units
R	*range* of a set of numbers, that is, the difference between the largest number and the smallest number	range of the n observed values in a subgroup (sample) (the symbol R is also used to designate the moving range, in Section 30)
s	*sample standard deviation*	standard deviation of the n observed values in a subgroup (sample) $s = \sqrt{\frac{(X_1 - \bar{X})^2 + \cdots + (X_n - \bar{X})^2}{n - 1}}$ or expressed in a form more convenient for computation purposes $s = \sqrt{\frac{n(X_1{}^2 + \cdots + X_n{}^2) - (X_1 + \cdots + X_n)^2}{n(n - 1)}}$
u		*defects per unit*, the number of defects in a sample of n units divided by n
X	observed value of a measurable characteristic; specific observed values are	

Symbol	General	In Part 3, Control Charts
	designated X_1, X_2, X_3, etc.; also used to designate a measurable characteristic	
$\overline{X}$	*average* (arithmetic mean); the sum of the n observed values divided by n	average of the n observed values in a subgroup (sample) $\overline{X} = \frac{X_1 + X_2 + \cdots + X_n}{n}$

Qualified Symbols

$\sigma_{\overline{X}}'$, σ_s', σ_R', σ_p'; etc.	standard deviation of values of $\overline{X}$, s, R, p, etc.	standard deviation of the sampling distribution of $\overline{X}$, s, R, p, etc.
$\overline{\overline{X}}$, $\bar{s}$, $\overline{R}$, $\bar{p}$, etc.	*average* of a set of values of X, s, R, p, etc. (the overbar notation signifies an average value)	average of the set of k subgroup (sample) values of $\overline{X}$, s, R, p, etc., under consideration; for samples of unequal size, an over-all or weighted average
$\overline{X}'$, σ', p', etc.	mean, standard deviation, fraction defective, etc., of the universe	
$\overline{X}_0'$, σ_0', p_0', etc.		standard value of $\overline{X}'$, σ', p', etc., adopted for computing control limits of a control chart for the case, Control-Standard Given (see Sections 18 to 27)

GENERAL PRINCIPLES

1. Purpose

Part 3 of the Manual gives formulas, tables, and examples which are useful in applying the *control chart* method [*1*][1] of analysis and presentation of data. This method requires that the data be obtained from several samples or that the data be capable of subdivision into subgroups on the basis of relevant engineering information. Although the principles of this Part are applicable generally to many kinds of data, they will be discussed herein largely in terms of the quality of materials and manufactured products.

The control chart method provides a criterion for detecting lack of statistical control. Lack of statistical control in data indicates that ob-

[1] The italic numbers in brackets refer to the list of references appended to Part 3.

served variations in quality are greater than should be left to chance. Freedom from indications of lack of control is desirable for scientific evaluation of data and the determination of quality.

The control chart method lays emphasis on the *order* or grouping of the observations in a set, with respect to time, place, source, or any other consideration that provides a basis for a classification which may be of significance in terms of known conditions under which the observations were obtained.

This concept of order is illustrated by the data in Table 1 in which the width in inches to the nearest 0.0001 in. is given for 60 specimens of Grade BB zinc which were used in ASTM atmospheric corrosion tests. At the left of the table, the data are tabulated without regard to relevant information. At the right they are shown arranged in ten subgroups, where each subgroup relates to the specimens from a separate milling. The information regarding origin is relevant engineering information which makes it possible to apply the control chart method to these data.

See Example 3, p. 102.

TABLE 1—*Comparison of data before and after subgrouping (width in inches of specimens of Grade BB zinc).*

Before Subgrouping

0.5005	0.5005	0.4996
0.5000	0.5002	0.4997
0.5008	0.5003	0.4993
0.5000	0.5004	0.4994
0.5005	0.5000	0.4999
0.5000	0.5005	0.4996
0.4998	0.5008	0.4996
0.4997	0.5007	0.4997
0.4998	0.5008	0.4995
0.4994	0.5010	0.4995
0.4999	0.5008	0.4997
0.4998	0.5009	0.4992
0.4995	0.5010	0.4995
0.4995	0.5005	0.4992
0.4995	0.5006	0.4994
0.4995	0.5009	0.4998
0.4995	0.5000	0.5000
0.4996	0.5001	0.4990
0.4998	0.5002	0.5000
0.5005	0.4995	0.5000

After Subgrouping

Subgroup (Milling)	Specimen 1	2	3	4	5	6
1	0.5005	0.5000	0.5008	0.5000	0.5005	0.5000
2	0.4998	0.4997	0.4998	0.4994	0.4999	0.4998
3	0.4995	0.4995	0.4995	0.4995	0.4995	0.4996
4	0.4998	0.5005	0.5005	0.5002	0.5003	0.5004
5	0.5000	0.5005	0.5008	0.5007	0.5008	0.5010
6	0.5008	0.5009	0.5010	0.5005	0.5006	0.5009
7	0.5000	0.5001	0.5002	0.4995	0.4996	0.4997
8	0.4993	0.4994	0.4999	0.4996	0.4996	0.4997
9	0.4995	0.4995	0.4997	0.4992	0.4995	0.4992
10	0.4994	0.4998	0.5000	0.4990	0.5000	0.5000

2. Terminology and Technical Background

Variation in quality from one unit of product to another is usually due to a very large number of causes. Those which it is possible to identify are termed *assignable causes*. Lack of control indicates one or more assignable causes. The vast majority of causes of variation may be inconsequential and cannot be identified. These are termed *chance causes*. However, causes of large variations in quality generally admit of ready identification.

In more detail we may say that for a constant system of chance causes, the averages, $\bar{X}$, the standard deviations, s, the values of fraction defective, p, or any other functions of the observations of a series of samples will exhibit statistical stability of the kind that may be expected in random samples from homogeneous material. The criterion of the quality control chart is derived from laws of chance variations for such samples, and failure to satisfy this criterion is taken as evidence of the presence of an assignable cause of variation.

As applied by the manufacturer to inspection data, the control chart provides a basis for *action*. Continued use of the control chart and the elimination of assignable causes as their presence is disclosed by failures to meet its criteria tend to reduce variability and to stabilize quality at aimed-at levels [*2–8*]. While the control chart method has been devised primarily for this purpose, it provides simple techniques and criteria that have been found useful in analyzing and interpreting other types of data as well.

3. Two Uses

The control chart method of analysis is used for the following two distinct purposes.

(*A*) *Control—No Standard Given*

To discover whether observed values of $\bar{X}$, s, p, etc., for several samples of n observations each, *vary among themselves* by an amount greater than should be attributed to chance. Control charts based entirely on the data from samples are used for detecting *lack of constancy* of the cause system.

(*B*) *Control with Respect to a Given Standard*

To discover whether observed values of $\bar{X}$, s, p, etc., for samples of n observations each, differ from standard values, $\bar{X}_0'$, σ_0', p_0', etc., by an amount greater than should be attributed to chance. The standard value may be an experience value based on representative prior data, or an economic value established on consideration of needs of service and cost of production, or a desired or aimed-at value designated by specification. It should be noted particularly that the standard value of σ', which is used not only for setting up control charts for s or R but also for computing control limits on control charts for $\overline{X}$, should almost invariably be an experience value based on representative *prior* data. Control charts based on such standards are used particularly in inspection to control processes and to maintain quality uniformly at the level desired.

4. Breaking up Data into Rational Subgroups

One of the essential features of the control chart method is what is referred to as breaking up the data into rationally chosen subgroups called "rational subgroups"; that is, classifying the observations under consideration into subgroups or samples, *within* which the variations may be considered on engineering grounds to be due to nonassignable chance causes only, but *between* which the differences may be due to assignable causes whose presence is suspected or considered possible.

This part of the problem depends on technical knowledge and familiarity with the conditions under which the material sampled was produced and the conditions under which the data were taken. By identifying each sample with a time or a source, specific causes of trouble may be more readily traced and corrected, if advantageous and economical. Inspection and test records, giving observations in the order in which they were taken, provide directly a basis for subgrouping with respect to time. This is commonly advantageous in manufacture where it is important, from the standpoint of quality, to maintain the production cause system constant with time.

It should always be remembered that analysis will be greatly facilitated if, when planning for the collection of data in the first place, care is taken to so select the samples that the data from each sample can properly be treated as a separate rational subgroup, and that the samples are identified in such a way as to make this possible.

5. General Technique in Using Control Chart Method

The general technique (see Ref *1*, Criterion I, Chapter XX) of the control chart method is as follows.

Given a set of observations, to determine whether an assignable cause of variation is present:

(*a*) Classify the total number of observations into k rational subgroups (samples) having $n_1, n_2, \ldots, n_k$ observations, respectively. Make subgroups of *equal size*, if practicable. It is usually preferable to make subgroups not smaller than $n = 4$.

(*b*) For each statistic ($\overline{X}, s, R, p$, etc.) to be used, construct a control chart with control limits in the manner indicated in the subsequent sections.

(*c*) If one or more of the observed values of $\overline{X}, s, R, p$, etc., for the k subgroups (samples) fall outside the control limits, take this fact as an indication of the presence of an assignable cause.

6. Control Limits

In both uses indicated in Section 3, the control chart consists essentially of symmetrical limits (control limits) placed above and below a central line. The central line in each case indicates the expected or average value of $\overline{X}$, s, R, p, etc. for subgroups (samples) of n observations each.

The control limits here used, referred to as "3-sigma control limits," are placed at a distance of three standard deviations from the central line, where by standard deviation is meant the standard deviation of the sampling distribution of the statistical measure in question ($\overline{X}$, s, R, p, etc.) for subgroups (samples) of size n. Note that this standard deviation is *not* the computed standard deviation of the subgroup values (of $\overline{X}$, s, R, p, etc.) plotted on the chart, but is computed from the variations within the subgroups (see Supplement B, Note 1).

Throughout this part of the Manual such standard deviations of the sampling distributions will be designated as $\sigma_{\overline{X}}'$, σ_s', σ_R', σ_p', etc., and these symbols, which consist of σ' and a subscript, will be used only in this restricted sense.

For measurement data, if $\overline{X}'$ and σ' are known, we have

control limits for averages	(expected $\overline{X}$) $\pm\ 3\sigma_{\overline{X}}'$
standard deviations	(expected s) $\pm\ 3\sigma_s'$
ranges	(expected R) $\pm\ 3\sigma_R'$

where the various expected values are derived from $\overline{X}'$ or σ'. For attribute data, if p' is known, we have

control limits for values of p	(expected p) $\pm\ 3\sigma_p'$

where expected $p = p'$.

Figure 1 indicates the features of a control chart for averages.

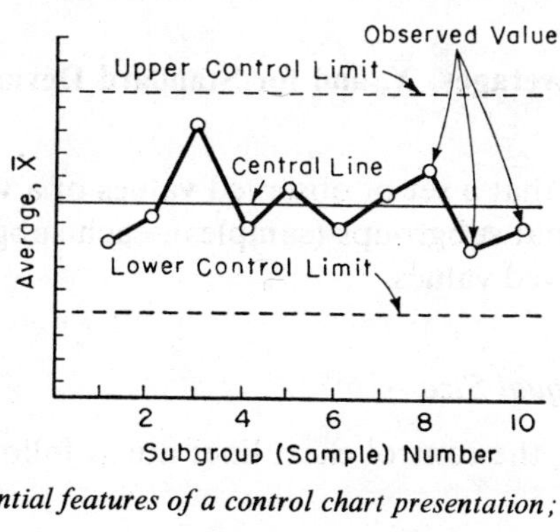

FIG. 1—*Essential features of a control chart presentation; chart for averages.*

The choice of the factor 3 (a multiple of the expected standard deviation of $\bar{X}$, s, R, p, etc.) in these limits is an economic choice based on experience that covers a wide range of industrial applications of the control chart, rather than on any exact value of probability (see Supplement B, Note 2). It is one that has proved satisfactory for use as a criterion for *action*, that is, for looking for assignable causes of variation.

CONTROL—NO STANDARD GIVEN

7. Introduction

Sections 7 to 17 cover the technique of analysis for control when no standard is given, as noted under (A) in Section 2. Here standard values of $\bar{X}'$, σ', p', etc., are *not given*, hence values derived from the numerical observations are used in arriving at central lines and control limits. This is the situation that exists when the problem at hand is the analysis and presentation of a given set of experimental data. This situation is also met in the initial stages of a program using the control chart method for controlling quality during production. Available information regarding the quality level and variability resides in the data *to be analyzed* and the central lines and control limits are based on values derived from those data. For a contrasting situation, see Section 18.

For a set of data analyzed by the control chart method, when may a state of control be assumed to exist? Assuming subgrouping based on time, it is usually not safe to assume that a state of control exists unless the plotted points for the following number of consecutive subgroups fall within 3-sigma control limits: (*a*) 25 for variables with small subgroups, say of 4 or 5, or (*b*) 15, 20, or 25 for attributes, when the expected number of defectives or of defects per subgroup is 1 to 4, over 4 to 7, or over 7, respectively. On the other hand, lack of control may be assumed to exist if one or more points fall outside the control limits in a much smaller number of subgroups, even 4 or 5.

8. Control Charts for Averages, $\bar{X}$, and for Standard Deviations, s—Large Samples

This section assumes that a set of observed values of a variable X can be subdivided into k rational subgroups (samples), each subgroup containing n = *more than* 25 observed values.

(*a*) *Large Samples of Equal Size*

For samples of size n, the control chart lines are as follows.

Formulas for control chart lines.

	Central Line	Control Limits	
For averages, $\bar{X}$	$\bar{\bar{X}}$	$\bar{\bar{X}} \pm 3\frac{\bar{s}}{\sqrt{n}}$	(1)
For standard deviations, s	$\bar{s}$	$\bar{s} \pm 3\frac{\bar{s}}{\sqrt{2(n-1)}}$	(2)

where

$$\bar{\bar{X}} = \text{the grand average of the observed values of } X \text{ for } all \text{ samples,}$$

$$= (\bar{X}_1 + \bar{X}_2 + \cdots + \bar{X}_k)/k \tag{3}$$

$$\bar{s} = \text{the average subgroup standard deviation,}$$

$$= (s_1 + s_2 + \cdots + s_k)/k \tag{4}$$

where the subscripts 1, 2, ..., k refer to the k subgroups, respectively, all of size n. (For a discussion of this formula, see Supplement B, Note 3.)

See Example 1, p. 100.

(*b*) *Large Samples of Unequal Size*

Use Formulas 1 and 2 but compute $\bar{\bar{X}}$ and $\bar{s}$ as follows

$$\bar{\bar{X}} = \text{the grand average of the observed values of } X \text{ for } all \text{ samples,}$$

$$= \frac{n_1\bar{X}_1 + n_2\bar{X}_2 + \cdots + n_k\bar{X}_k}{n_1 + n_2 + \cdots + n_k} \tag{5}$$

$$= \text{grand total of } X \text{ values divided by their total number,}$$

$$\bar{s} = \text{the weighted average standard deviation,}$$

$$= \frac{n_1 s_1 + n_2 s_2 + \cdots + n_k s_k}{n_1 + n_2 + \cdots + n_k} \tag{6}$$

where the subscripts 1, 2, ..., k refer to the k subgroups, respectively. (For a discussion of this formula, see Supplement B, Note 3.)

Then compute control limits for each sample size separately, using the individual sample size, n, in the formula for control limits.

See Example 2, p. 102.

When most of the samples are of approximately equal size, computing and plotting effort can be saved by the procedure given in Supplement B, Note 4.

9. Control Charts for Averages, $\bar{X}$, and for Standard Deviations, s—Small Samples

This section assumes that a set of observed values of a variable X is subdivided into k rational subgroups (samples), each subgroup containing $n = 25$ *or less* observed values.

(a) *Small Samples of Equal Size*

For samples of size n, the control chart lines are as follows.

Formulas for control chart lines.

	Central Line	Control Limits: Simplified Formula Using Factors in Table 2	Control Limits: Basic Formula	
For averages, $\bar{X}$	$\bar{\bar{X}}$	$\bar{\bar{X}} \pm A_3\bar{s}$	$\bar{\bar{X}} \pm 3\dfrac{\bar{s}}{c_4\sqrt{n}}$	(7)
For standard deviations, s	$\bar{s}$	$B_4\bar{s}$ and $B_3\bar{s}$	$\bar{s} \pm 3\dfrac{\bar{s}}{c_4\sqrt{2(n-1)}}$.	(8)

where

$\bar{\bar{X}}$ = the grand average of observed values of X for *all* samples,

$$\bar{s} = \frac{s_1 + s_2 + \cdots + s_k}{k} \tag{9}$$

where s_1, s_2, etc., refer to the observed standard deviations for the first, second, etc., samples and factors c_4, A_3, B_3, and B_4 are given in Table 2.

See Example 3, p. 102.

Formula 8 is an approximate formula suitable for most practical purposes. The values of B_3 and B_4 given in the tables are computed from the exact equation in Supplement A (Eqs 46 and 47). For a discussion of Eq 9, see Supplement B, Note 3.

(b) *Small Samples of Unequal Size*

For *small samples of unequal size*, use Formulas 7 and 8 (or corresponding factors) for computing control chart lines. Compute $\bar{\bar{X}}$ by Eq 5. Obtain separate derived values of $\bar{s}$ for the different sample sizes by the following working rule: Compute σ_e, the over-all *average* value of the ratio observed s/c_4 for the individual samples; then compute $\bar{s} = c_4\sigma_e$ for each sample size n. As shown in Example 4, the computation can be simplified by combining in separate groups all samples having the same sample size n. Control limits may then be determined separately for each sample size. These difficulties can be avoided by planning the collection of data so that the samples are made of equal size.

See Example 4, p. 105.

10. Control Charts for Averages, $\bar{X}$, and for Ranges, R—Small Samples

This section assumes that a set of observed values of a variable X is subdivided into k rational subgroups (samples), each subgroup containing $n = 10$ *or less* observed values.

The range, R, of a sample is the difference between the largest observation and the smallest observation. When $n = 10$ or less, simplicity and economy of effort can be obtained by using control charts for $\bar{X}$ and R in place of control charts for $\bar{X}$ and s. The range is not recommended, however, for samples of more than 10 observations, since it becomes rapidly less effective than the standard deviation as a detector of assignable causes as n increases beyond this value. In some circumstances it may be found satisfactory to use the control chart for ranges for samples up to $n = 15$, as when data are plentiful or cheap. On occasion it may be desirable to use the chart for ranges for even larger samples; for this reason Table 2 gives factors for samples as large as $n = 25$.

(*a*) *Small Samples of Equal Size*

For samples of size n, the control chart lines are as follows.

Formulas for control chart lines.

	Central Line	Control Limits: Simplified Formula Using Factors in Table 2	Control Limits: Basic Formula	
For averages, $\bar{X}$	$\bar{\bar{X}}$	$\bar{\bar{X}} \pm A_2\bar{R}$	$\bar{\bar{X}} \pm 3\dfrac{\bar{R}}{d_2\sqrt{n}}$	(10)
For ranges, R	$\bar{R}$	$D_4\bar{R}$ and $D_3\bar{R}$	$\bar{R} \pm 3\dfrac{d_3\bar{R}}{d_2}$	(11)

where

$\bar{X}$ = the grand average of observed values of X for *all* samples,

R = the average value of range R for the k individual samples,

$$= (R_1 + R_2 + \cdots + R_k)/k \tag{12}$$

and factors d_2, A_2, D_3 and D_4 are given in Table 2 and d_3 in Table 27.

See Example 5, p. 104.

(b) Small Samples of Unequal Size

For *small samples* of *unequal size*, use Formulas 10 and 11 (or corresponding factors) for computing control chart lines. Compute $\bar{X}$ by Eq 5. Obtain separate derived values of $\bar{R}$ for the different sample sizes by the following working rule: compute σ_e the over-all *average* value of the ratio observed R/d_2 for the individual samples then compute $\bar{R} = d_2\sigma_e$ for each sample size n. As shown in Example 6, the computation can be simplified by combining in separate groups all samples having the same sample size n. Control limits may then be determined separately for each sample size. These difficulties can be avoided by planning the collection of data so that the samples are made of equal size.

See Example 6, p. 105.

Formulas for central lines and control limits.[a]

	Central Line	Control Limits
Averages using s	$\bar{X}$	$\bar{X} \pm A_3\bar{s}$ ($\bar{s}$ as given by Eq 9)
Averages using R	$\bar{X}$	$\bar{X} \pm A_2\bar{R}$ ($\bar{R}$ as given by Eq 12)
Standard deviations	$\bar{s}$	$B_4\bar{s}$ and $B_3\bar{s}$ ($\bar{s}$ as given by Eq 9)
Ranges	$\bar{R}$	$D_4\bar{R}$ and $D_3\bar{R}$ ($\bar{R}$ as given by Eq 12)

[a] Control—no standard given ($\bar{X}'$, σ', not given)—small samples of equal size.

11. Summary, Control Charts for $\bar{X}$, s, and R—No Standard Given

The most useful formulas and equations from Sections 7 to 10, inclusive, are collected and are followed by Table 2 which gives the factors used in these and other formulas.

12. Control Charts for Attributes Data

Although in what follows the function p is designated "fraction defective," the methods described can be applied quite generally and p may in fact be used to represent the ratio of the number of items, occurrences, etc. that

TABLE 2—*Factors for computing control chart lines—no standard given.*[a]

Observations in Sample, n	Chart for Averages: Factors for Control Limits		Chart for Standard Deviations: Factors for Central Line	Factors for Control Limits		Chart for Ranges: Factors for Central Line	Factors for Control Limits	
	A_2	A_3	c_4	B_3	B_4	d_2	D_3	D_4
2	1.880	2.659	0.7979	0	3.267	1.128	0	3.267
3	1.023	1.954	0.8862	0	2.568	1.693	0	2.575
4	0.729	1.628	0.9213	0	2.266	2.059	0	2.282
5	0.577	1.427	0.9400	0	2.089	2.326	0	2.115
6	0.483	1.287	0.9515	0.030	1.970	2.534	0	2.004
7	0.419	1.182	0.9594	0.118	1.882	2.704	0.076	1.924
8	0.373	1.099	0.9650	0.185	1.815	2.847	0.136	1.864
9	0.337	1.032	0.9693	0.239	1.761	2.970	0.184	1.816
10	0.308	0.975	0.9727	0.284	1.716	3.078	0.223	1.777
11	0.285	0.927	0.9754	0.321	1.679	3.173	0.256	1.744
12	0.266	0.886	0.9776	0.354	1.646	3.258	0.284	1.716
13	0.249	0.850	0.9794	0.382	1.618	3.336	0.308	1.692
14	0.235	0.817	0.9810	0.406	1.594	3.407	0.329	1.671
15	0.223	0.789	0.9823	0.428	1.572	3.472	0.348	1.652
16	0.212	0.763	0.9835	0.448	1.552	3.532	0.364	1.636
17	0.203	0.739	0.9845	0.466	1.534	3.588	0.379	1.621
18	0.194	0.718	0.9854	0.482	1.518	3.640	0.392	1.608
19	0.187	0.698	0.9862	0.497	1.503	3.689	0.404	1.596
20	0.180	0.680	0.9869	0.510	1.490	3.735	0.414	1.586
21	0.173	0.663	0.9876	0.523	1.477	3.778	0.425	1.575
22	0.167	0.647	0.9882	0.534	1.466	3.819	0.434	1.566
23	0.162	0.633	0.9887	0.545	1.455	3.858	0.443	1.557
24	0.157	0.619	0.9892	0.555	1.445	3.895	0.452	1.548
25	0.153	0.606	0.9896	0.565	1.435	3.931	0.459	1.541
Over 25	...	...	...	...[b]	...[c]	...	...	...

[a] The convenient tabular arrangement of this table and that of Table 3 corresponds closely to that used by Pearson [4]. Table 2 was revised in 1975 by deleting old columns A_1 and c_2 and adding new columns A_3 and c_4; see Table 27.

[b] $1 - 3/\sqrt{2(n-1)}$.

[c] $1 + 3/\sqrt{2(n-1)}$.

possess some given attribute to the total number of items under consideration.

The fraction defective, p, is particularly useful in analyzing inspection and test results that are obtained on a "go no-go" basis (method of attributes). Also it is used in analyzing results of measurements that are made on a scale and recorded (method of variables). In the latter case, p may be used to represent the fraction of the total number of measured values falling above any limit, below any limit, between any two limits, or outside any two limits.

The function p is used widely to represent the "fraction defective," that is, the ratio of the number of defective units (articles, parts, specimens, etc.) to the total number of units under consideration. The fraction defective is used as a measure of quality with respect to a single quality characteristic or with respect to two or more quality characteristics treated collectively. In this connection it is important to distinguish between a "defect" and a "defective." A "defect" is a *single* instance of a failure to meet some requirement, such as a failure to comply with a particular requirement imposed on a unit of product with respect to a single quality characteristic. For example, a unit containing departures from requirements of the drawings and specifications with respect to (*1*) a particular dimension, (*2*) finish, and (*3*) absence of chamfer, contains three defects. The word "defective" is here used as a noun and is defined as a defective unit (article, part, specimen, etc.), that is, a unit containing one or more "defects" with respect to the quality characteristic under consideration.

When only a single quality characteristic is under consideration, and when only one defect can occur on a unit, the number of defectives in a sample will equal the number of defects in that sample. However, it is suggested that under these circumstances the phrase "number of defectives" be used rather than "number of defects."

13. Control Chart for Fraction Defective, p

This section assumes that the total number of units tested is subdivided into k rational subgroups (samples) containing of $n_1, n_2, \ldots, n_k$ units, respectively, for each of which a value of p is computed.

Ordinarily the control chart for p is most useful when the samples are large, say when n is greater than 50 to 100; more specifically, when the expected number of defective units (or other occurrences of interest) per sample is four or more, that is, the expected pn is four or more.

The average fraction defective, $\bar{p}$, is defined as

$$\bar{p} = \frac{\text{total number of defectives in all samples}}{\text{total number of units in all samples}} \tag{13}$$

$$= \text{fraction defective in the complete set of test results.}$$

(*a*) *Samples of Equal Size*

For samples of size n, the control chart lines are as follows.

Formulas for control chart lines.

	Central Line	Control Limits	
For values of p	$\bar{p}$	$\bar{p} \pm 3\sqrt{\frac{\bar{p}(1-\bar{p})}{n}}$	(14)

See Example 7, p. 105.

When $\bar{p}$ is small, say less than 0.10, the factor $1 - \bar{p}$ may be replaced by unity for most practical purposes, which gives the simple relation

$$\text{control limits for } p = \bar{p} \pm 3\sqrt{\frac{\bar{p}}{n}} \tag{14a}$$

(b) *Samples of Unequal Size*

Proceed as for samples of equal size but compute control limits for each sample size separately.

When the data are in the form of a series of k subgroup values of p and the corresponding sample sizes n, $\bar{p}$ may be computed conveniently by the relation

$$\bar{p} = \frac{n_1p_1 + n_2p_2 + \cdots + n_kp_k}{n_1 + n_2 + \cdots + n_k} \tag{15}$$

where the subscripts 1, 2, ..., k refer to the k subgroups.

When most of the samples are of approximately equal size, computation and plotting effort can be saved by the procedure given in Supplement B, Note 4.

See Example 8, p. 107.

14. Control Chart for Number of Defectives, *pn*

The control chart for pn, number of defectives in a sample of size n, is the equivalent of the control chart for p, for which it is a convenient practical substitute when all samples have the same size, n. It makes direct use of the number of defectives, pn, in a sample (pn = the fraction defective times the sample size.)

For samples of size n, the control chart lines are as follows.

Formulas for control chart lines.

	Central Line	Control Limits	
For values of pn	$\bar{p}n$	$\bar{p}n \pm 3\sqrt{\bar{p}n(1-\bar{p})}$	(16)

where

$$\bar{p}n = \text{total number of defectives in all samples/number of samples} \quad (17)$$

$$= \text{the } \textit{average} \text{ number of defectives in the } k \text{ individual samples, and}$$

$\bar{p}$ = the value given by Eq 13.

When $\bar{p}$ is small, say less than 0.10, the factor $1 - \bar{p}$ may be replaced by unity for most practical purposes, which gives the simple relation

$$\text{control limits for } pn = \bar{p}n \pm 3\sqrt{\bar{p}n}$$

$$= \text{(average number of defectives)} \quad (18)$$

$$\pm 3\sqrt{\text{(average number of defectives)}}$$

where "average number of defectives" means *average number* in samples of equal size.

See Example 7, p. 105.

When the sample size, n, varies from sample to sample, the control chart for p (Section 13) is recommended in preference to the control chart for pn; in this case, a graphical presentation of values of pn does not give an easily understood picture, since the expected values, $\bar{p}n$, (central line on the chart) varies with n, and therefore the plotted values of pn become more difficult to compare.

When only a single quality characteristic is under consideration, and when only one defect can occur on a unit, the word "defect" can be substituted for the word "defective" throughout the discussion of this section but this practice is not recommended.

15. Control Chart for Defects per Unit, *u*

In inspection and testing, there are circumstances where it is possible for several defects to occur on a single unit (article, part, specimen, unit length, unit area, etc.) of product, and it is desired to control the number of defects per unit, rather than the fraction defective. For any given sample of units, the numerical value of defects per unit, u, is equal to the number of defects in all the units in the sample divided by the number of units in the sample.

The control chart for u, defects per unit in a sample, is convenient for a product composed of units for which inspection covers more than one characteristic, such as dimensions checked by gages, electrical and mechanical characteristics checked by tests, and visual defects observed by the eye. Under these circumstances several independent defects may occur on one unit of product and a better measure of quality is obtained by making a count of all defects observed and dividing by the number of units inspected to give a value of defects per unit, rather than by merely counting the number of defective units to give a value of fraction defective. This is particularly the case for complex assemblies where the occurrence of two or more defects on a unit may be relatively frequent. However, only *independent* defects are counted. Thus, if two defects occur on one unit of product and the second is caused by the first, only the first is counted.

The control chart for defects per unit (more especially the chart for number of defects, see Section 16) is a particularly convenient one to use when the number of possible defects on a unit is indeterminate, as for physical defects (finish or surface irregularities, flaws, pinholes, etc.) on such products as textiles, wire, sheet materials, etc., which are continuous or extensive. Here, the opportunity for defects may be numerous though the chances of a defect occurring at any one spot may be small.

This section assumes that the total number of units tested is subdivided into k rational subgroups (samples) consisting of $n_1, n_2, \ldots, n_k$ units, respectively, for each of which a value of u is computed.

The average defects per unit, $\bar{u}$, is defined as

$$\bar{u} = \frac{\text{total number of defects in all samples}}{\text{total number of units in all samples}}$$
$$= \text{defects per unit in the complete set of test results.} \qquad (19)$$

The simplified relations shown for control limits for defects per unit assume that for each of the characteristics under consideration the ratio of the expected number of defects to the possible number of defects is small, say less than 0.10, an assumption that is commonly satisfied in quality control work. For an explanation of the nature of distribution involved, see Supplement B, Note 5.

(*a*) *Samples of Equal Size*

For samples of size n (n = number of units), the control chart lines are as follows.

Formulas for control chart lines.

	Central Line	Control Limits	
For values of u	$\bar{u}$	$\bar{u} \pm 3\sqrt{\dfrac{\bar{u}}{n}}$	(20)

For samples of equal size, a chart for number of defects is recommended, see Section 16. In the special case where each sample consists of only one unit, that is, $n = 1$, then the chart for u (defects per unit) is identical with that chart for c (number of defects) and may be handled in accordance with

Section 16. In this case the chart may be referred to either as a chart for defects per unit or as a chart for number of defects, but the latter designation is recommended.

See Example 9, p. 109.

(b) Samples of Unequal Size

Proceed as for samples of equal size but compute the control limits for each sample size separately.

When the data are in the form of a series of subgroup values of u and the corresponding sample sizes, $\bar{u}$ may be computed by the relation

$$\bar{u} = \frac{n_1 u_1 + n_2 u_2 + \cdots + n_k u_k}{n_1 + n_2 + \cdots + n_k} \tag{21}$$

where as before, the subscripts 1, 2, . . ., k refer to the k subgroups.

Note that n_1, n_2, etc., need not be whole numbers. For example, if u represents defects per 1000 ft of wire, samples of 4000 ft, 5280 ft, etc., constitute 4.0, 5.28, etc., units of 1000 ft.

When most of the samples are of approximately equal size, computing and plotting effort can be saved by the procedure in Supplement B, Note 4.

See Example 10, p. 109.

16. Control Chart for Number of Defects, *c*

The control chart for c, the number of defects in a sample, is the equivalent of the control chart for u, for which it is a convenient practical substitute when all samples have the *same size n* (number of units).

(a) Samples of Equal Size

For samples of equal size, if the average number of defects per sample is $\bar{c}$, the control chart lines are as follows.

Formulas for control chart lines.

	Central Line	Control Limits	
For values of c	$\bar{c}$	$\bar{c} \pm 3\sqrt{\bar{c}}$	(22)

where

$$\bar{c} = \frac{\text{total number of defects in all samples}}{\text{number of samples}} \tag{23}$$

$$= \text{average number of defects per sample.}$$

The use of c is especially convenient when there is no natural unit of product, as for defects over a surface or along a length, and where the problem is to determine uniformity of quality in equal lengths, areas, etc., of product.

See Example 9, p. 109, and Example 11, p. 111.

(b) Samples of Unequal Size

For samples of unequal size, first compute the average defects per unit $\bar{u}$, by Eq 19; then compute the control limits for each sample size separately as follows.

Formulas for control chart lines.

	Central Line	Control Limits	
For values of c	$\bar{u}n$	$\bar{u}n \pm 3\sqrt{\bar{u}n}$	(24)

The control chart for u is recommended as preferable to the control chart for c when the sample size varies from sample to sample for reasons stated in discussing the control charts for p and pn.

17. Summary, Control Charts for *p*, *pn*, *u*, and *c*—No Standard Given

The formulas of Sections 13 to 16, inclusive, are collected as follows for convenient reference.

Formulas for central lines and control limits.

Control—No Standard Given—Attributes Data			
	Central Line	Control Limits	Approximation
Fraction defective, p	$\bar{p}$	$\bar{p} \pm 3\sqrt{\dfrac{\bar{p}(1-\bar{p})}{n}}$	$\bar{p} \pm 3\sqrt{\dfrac{\bar{p}}{n}}$
Number of defectives, pn	$\bar{p}n$	$\bar{p}n \pm 3\sqrt{\bar{p}n(1-\bar{p})}$	$\bar{p}n \pm 3\sqrt{\bar{p}n}$
Defects per unit, u	$\bar{u}$	$\bar{u} \pm 3\sqrt{\dfrac{\bar{u}}{n}}$	
Number of defects, c			
samples of equal size	$\bar{c}$	$\bar{c} \pm 3\sqrt{\bar{c}}$	
samples of unequal size	$\bar{u}n$	$\bar{u}n \pm 3\sqrt{\bar{u}n}$	

CONTROL WITH RESPECT TO A GIVEN STANDARD

18. Introduction

Sections 19 to 27 cover the technique of analysis for control with respect to a given standard, as noted under (B) in Section 3. Here, standard values of $\overline{X}'$, σ', p', etc., are *given*, and are those corresponding to a given standard distribution. These standard values, designated as $\overline{X}_0'$, σ_0', p_0', etc., are used in calculating both central lines and control limits. (When only $\overline{X}_0'$ is given and no prior data are available for establishing a value of σ_0', analyze data from the first production period as in Sections 7 to 10, but use $\overline{X}_0'$ as the central line.)

Such standard values are usually based on a control chart analysis of previous data (for details, see Supplement B, Note 6), but may be based on the requirements as to quality level and variability which the product must meet or on some estimated or claimed capabilities of the production process. Note that these standard values are set up *before* the detailed analysis of the data at hand is undertaken and frequently before the data to be analyzed are collected. In addition to the standard values, only the information regarding sample size or sizes is required in order to compute central lines and control limits.

For example, the values to be used as central lines on the control charts are

for averages,	$\overline{X}_0'$
for standard deviations,	$c_4\sigma_0'$
for ranges,	$d_2\sigma_0'$
for values of p,	p_0'
etc.,	

where factors c_4 and d_2, which depend only on the sample size, n, are given in Table 3, and defined in Supplement A.

Note that control with respect to a given standard may be a more exacting requirement than control with no standard given, described in Sections 7 to 17. The data must exhibit not only control but control at a standard level and with no more than standard variability.

Where the control limits obtained in the analysis of a set of data are extended into the future and used as a basis for purposive control of quality during production, this is the equivalent of adopting, as standard, values found from the analysis of the prior data. Standard values so obtained may be tentative and subject to revision as more experience is accumulated (for details, see Supplement B, Note 6).

NOTE—Two situations not covered specifically herein should be mentioned.

(*a*) In some cases a standard value of $\overline{X}'$ is arbitrarily set by specification or otherwise, but no standard value is given for σ'. Here σ' is estimated from the analysis of the data at hand

and the problem is essentially one of controlling $\bar{X}$ at the standard level $\bar{X}_0'$ that has been arbitrarily set.

(*b*) In other cases, interest centers on controlling conformance to specified minimum and maximum limits within which material is considered acceptable, sometimes established without regard to the actual variation experienced in production. Such limits may prove unrealistic when data are accumulated and an estimate of the standard deviation, say σ^*, of the process is obtained therefrom. If the natural spread of the process (a band having a width of $6\sigma^*$), is wider than the spread between the specified limits, it is necessary either to adjust the specified limits or to operate within a band narrower than the process capability. Conversely, if the spread of the process is narrower than the spread between the specified limits, the process will deliver a more uniform product than required. In the latter event when only maximum and minimum limits are specified, the process can be operated at a level above or below the indicated mid-value without risking the production of significant amounts of unacceptable material.

19. Control Charts for Averages, $\bar{X}$, and for Standard Deviation, s

For samples of size n, the control chart lines are as follows.

Formula for control chart lines.

	Central Line	Control Limits: Simplified Formula Using Factors in Table 3	Control Limits: Basic Formula	
For averages, $\bar{X}$	$\bar{X}_0'$	$\bar{X}_0' \pm A\sigma_0'$	$\bar{X}_0' \pm 3\dfrac{\sigma_0'}{\sqrt{n}}$	(25)
For standard deviations, s	$c_4\sigma_0'$	$B_6\sigma_0'$ and $B_5\sigma_0'$	$c_4\sigma_0' \pm 3\dfrac{\sigma_0'}{\sqrt{2n}}$	(26)[a]

[a] Formula 26 is an approximate formula suitable for most practical purposes. The values of B_5 and B_6 given in the tables are computed from the exact equation in Supplement A (Eqs 46 and 47).

For samples of n greater than 25, consider $c_4 = 1$.

See Example 12, p. 113.

For samples of $n = 25$ or less, use Table 3 for factors A, B_5, and B_6. Factors c_4, A, B_5 and B_6 are defined in Supplement A.

See Example 13, p. 114; Example 14, p. 115; and Example 15, p. 116.

20. Control Chart for Ranges, R

The range, R, of a sample is the difference between the largest observation and the smallest observation.

For samples of size n, the control chart lines are as follows.

Formulas for control chart lines.

	Central Line	Control Limits: Simplified Formula Using Factors in Table 3	Control Limits: Basic Formula	
For range, R	$d_2\sigma_0'$	$D_2\sigma_0'$ and $D_1\sigma_0'$	$d_2\sigma_0' \pm 3\,d_3\sigma_0'$	(27)

Use Table 3 for factors d_2, D_1, and D_2.
Factors d_2, d_3, D_1, and D_2 are defined in Supplement A.
For comments on the use of the control chart for ranges, see Section 10.
See Example 16, p. 117.

21. Summary, Control Charts for $\bar{X}$, s, and $\bar{R}$—Standard Given

The most useful formulas from Sections 19 and 20 are summarized as follows and are followed by Table 3 which gives the factors used in these and other formulas.

Formulas for central lines and control limits.

Control with Respect to a Given Standard ($\bar{X}_0'$, σ_0' given)		
	Central Line	Control Limits
Averages, $\bar{X}$	$\bar{X}_0'$	$\bar{X}_0' \pm A\sigma_0'$
Standard deviations, s	$c_4\sigma_0'$	$B_6\sigma_0'$ and $B_5\sigma_0'$
Ranges, R	$d_2\sigma_0'$	$D_2\sigma_0'$ and $D_1\sigma_0'$

22. Control Charts for Attributes Data

The definitions of terms and the discussions in Sections 12 to 16, inclusive, on the use of the fraction defective, p, number of defectives, pn, defects per unit, u, and number of defects, c, as measures of quality are equally applicable to the sections which follow and will not be repeated here. It will suffice to discuss the central lines and control limits when standards are given.

23. Control Chart for Fraction Defective, p

Ordinarily, the control chart for p is most useful when samples are large, say, when n is greater than 50 or 100; more specifically, when the expected number of defective units (or other occurrences of interest) per sample is four or more, that is, the expected pn = four or more.

Table 3—*Factors for computing control chart lines—standard given.*[a]

	Chart for Averages	Chart for Standard Deviations			Chart for Ranges		
	Factors for Control Limits	Factor for Central Line	Factors for Control Limits		Factor for Central Line	Factors for Control Limits	
Observations in Sample, n	A	c_4	B_5	B_6	d_2	D_1	D_2
2	2.121	0.7979	0	2.606	1.128	0	3.686
3	1.732	0.8862	0	2.276	1.693	0	4.358
4	1.500	0.9213	0	2.088	2.059	0	4.698
5	1.342	0.9400	0	1.964	2.326	0	4.918
6	1.225	0.9515	0.029	1.874	2.534	0	5.078
7	1.134	0.9594	0.113	1.806	2.704	0.204	5.204
8	1.061	0.9650	0.179	1.751	2.847	0.388	5.306
9	1.000	0.9693	0.232	1.707	2.970	0.547	5.393
10	0.949	0.9727	0.276	1.669	3.078	0.687	5.469
11	0.905	0.9754	0.313	1.637	3.173	0.811	5.535
12	0.866	0.9776	0.346	1.610	3.258	0.922	5.594
13	0.832	0.9794	0.374	1.585	3.336	1.025	5.647
14	0.802	0.9810	0.399	1.563	3.407	1.118	5.696
15	0.775	0.9823	0.421	1.544	3.472	1.203	5.741
16	0.750	0.9835	0.440	1.526	3.532	1.282	5.782
17	0.728	0.9845	0.458	1.511	3.588	1.356	5.820
18	0.707	0.9854	0.475	1.496	3.640	1.424	5.856
19	0.688	0.9862	0.490	1.483	3.689	1.487	5.891
20	0.671	0.9869	0.504	1.470	3.735	1.549	5.921
21	0.655	0.9876	0.516	1.459	3.778	1.605	5.951
22	0.640	0.9882	0.528	1.448	3.819	1.659	5.979
23	0.626	0.9887	0.539	1.438	3.858	1.710	6.006
24	0.612	0.9892	0.549	1.429	3.895	1.759	6.031
25	0.600	0.9896	0.559	1.420	3.931	1.806	6.056
Over 25	$3/\sqrt{n}$	...	...	...	...	...	...

[a] Table 3 was revised in 1975 by deleting old columns, c_2, B_1, and B_2, adding new columns c_4, B_5, and B_6 and revising columns D_1 and D_2. See Table 27.

For samples of size n, where p_0' is the standard value of p', the control chart lines are as follows.

Formulas for control chart lines.

	Central Line	Control Limits	
For values of p	p_0'	$p_0' \pm 3\sqrt{\dfrac{p_0'(1-p_0')}{n}}$	(28)

See Example 17, p. 118.

When p_0' is small, say less than 0.10, the factor $1 - p_0'$ may be replaced by unity for most practical purposes, which gives the simple relation

$$\text{control limits for } p = p_0' \pm 3\sqrt{\frac{p_0'}{n}} \tag{28a}$$

For samples of unequal size, proceed as for samples of equal size but compute control limits for each sample size separately.

See Example 18, p. 120.

When detailed inspection records are maintained, the control chart for p may be broken down into a number of component charts with advantage.

See Example 19, p. 121.

24. Control Chart for Number of Defectives, *pn*

The control chart for *pn*, number of defectives in a sample, is the equivalent of the control chart for fraction defective, *p*, for which it is a convenient practical substitute, particularly when all samples have the *same size*, ***n***. It makes direct use of the number of defectives, *pn*, in a sample (*pn* = the product of the fraction defective and the sample size).

See Example 17, p. 118.

For samples of size n, where p_0' is the standard value of p', the control chart lines are as follows.

Formulas for control chart lines.

	Central Line	Control Limits	
For values of *pn*	$p_0'n$	$p_0'n \pm 3\sqrt{p_0'n(1 - p_0')}$	(29)

When p_0' is small, say less than 0.10, the factor $1 - p_0'$ may be replaced by unity for most practical purposes, which gives the simple relation

$$\text{control limits for } pn = p_0'n \pm 3\sqrt{p_0'n} \tag{30}$$

As noted in Section 14, the control chart for p is recommended as preferable to the control chart for pn when the sample size varies from sample to sample.

When only a single quality characteristic is under consideration, and when only one defect can occur on a unit, the word "defect" can be substituted for the word "defective" throughout the discussion of this article but this practice is not recommended.

25. Control Chart for Defects per Unit, *u*

For samples of size n (n = number of units), where u_0' is the standard value of u', the control chart lines are as follows.

Formulas for control chart lines.

	Central Line	Control Limits	
For values of u	u_0'	$u' \pm 3\sqrt{\frac{u_0'}{n}}$	(31)

See Example 20, p. 125.

As noted in Section 15, the relations given here assume that for each of the characteristics under consideration, the ratio of the expected to the possible number of defects is small, say less than 0.10.

If u represents "defects per 1000 ft of wire," a "unit" is 1000 ft of wire. Then if a series of samples of 4000 ft are involved, u_0' represents the standard or expected number of defects per 1000 ft, and $n = 4$. Note that n need not be a whole number, for if samples comprise 5280 ft of wire each, $n = 5.28$, that is, 5.28 units of 1000 ft.

See Example 11, p. 111.

Where each sample consists of only one unit, that is $n = 1$, then the chart for u (defects per unit) is identical with the chart for c (number of defects) and may be handled in accordance with Section 26. In this case the chart may be referred to either as a chart for defects per unit or as a chart for number of defects, but the latter practice is recommended.

26. Control Chart for Number of Defects, *c*

The control chart for c, number of defects in a sample, is the equivalent of the control chart for defects per unit for which it is a convenient practical substitute when all samples have the *same size*, n (number of units). Here c is the *number of defects in a sample*.

If the standard value is expressed in terms of number of defects per sample of some given size, that is, expressed merely as c_0', and the samples are all of the same given size (same number of product units, same area of opportunity for defects, same sample length of wire, etc.), then the control chart lines are as follows.

Formulas for control chart lines (c_0' *given*).

	Central Line	Control Limits	
For number of defects, c	c_0'	$c_0' \pm 3\sqrt{c_0'}$	(32)

Use of c_0' is especially convenient when there is no natural unit of product, as for defects over a surface or along a length, and where the problem of interest is to compare uniformity of quality in samples of the same size, no matter how constituted.

See Example 21, p. 126.

When the sample size, n, (number of units) varies from sample to sample, and the standard value is expressed in terms of defects per unit, the control chart lines are as follows.

Formulas for control chart lines (u_0' *given*).

	Central Line	Control Limits	
For values of c	$u_0'n$	$u_0'n \pm 3\sqrt{u_0'n}$	(33)

Under these circumstances the control chart for u (Section 25) is recommended in preference to the control chart for c, for reasons stated in Section 14 in the discussion of control charts for p and for pn.

27. Summary, Control Charts for p, pn, u, and c—Standard Given

The formulas of Sections 22 to 26, inclusive, are collected as follows for convenient reference.

Formulas for central lines and control limits.

Control With Respect to a Given Standard (p_0', $p_0'n$, u_0' or c_0' given).			
	Central Line	Control Limits	Approximation
Fraction defective, p	p_0'	$p_0' \pm 3\sqrt{\dfrac{p_0'(1-p_0')}{n}}$	$p_0' \pm 3\sqrt{\dfrac{p_0'}{n}}$
Number of defectives, pn	$p_0'n$	$p_0'n \pm 3\sqrt{p_0'n(1-p_0')}$	$p_0'n \pm 3\sqrt{p_0'n}$
Defects per unit, u	u_0'	$u_0' \pm 3\sqrt{\dfrac{u_0'}{n}}$	
Number of defects, c			
samples of equal size (c_0' given)	c_0'	$c_0' \pm 3\sqrt{c_0'}$	
samples of unequal size (u_0' given)	$u_0'n$	$u_0'n \pm 3\sqrt{u_0'n}$	

CONTROL CHARTS FOR INDIVIDUALS

28. Introduction

Sections 28 to 30[2] deal with control charts for individuals, in which in-

[2] To be used with caution if the distribution of individual values is markedly asymmetrical.

dividual observations are plotted one by one. This type of control chart has been found useful more particularly in process control when only one observation is obtained per lot or batch of material or at periodic intervals from a process. This situation often arises when: (*a*) sampling or testing is destructive, (*b*) costly chemical analyses or physical tests are involved, and (*c*) the material sampled at any one time (such as a batch) is normally quite homogeneous, as for a well-mixed fluid or aggregate.

The purpose of such control charts is to discover whether the individual observed values differ from the expected value by an amount greater than should be attributed to chance.

When there is some definite rational basis for grouping the batches or observations into rational subgroups, as, for example, four successive batches in a single shift, the method shown in Section 29 may be followed. In this case the control chart for individuals is merely an adjunct to the more usual charts but will react more quickly to a sharp change in the process than the $\bar{X}$ chart. This may be important when a single batch represents a considerable sum of money.

When there is no definite basis for grouping the data, the control limits may be based on the variation between batches, as described in Section 30. A measure of this variation is obtained from moving ranges of two observations (the successive differences between individual observations which are arranged in chronological order).

A control chart for moving ranges may be prepared as a companion to the chart for individuals, if desired, using the formulas of Section 30. It should be noted that adjacent moving ranges are correlated, as they have one observation in common.

The methods of Sections 29 and 30 may be applied appropriately in some cases where more than one observation is obtained per lot or batch, as for example with very homogeneous batches of materials, for instance, chemical solutions, batches of thoroughly mixed bulk materials, etc., for which repeated measurements on a single batch show the within-batch variation (variation of quality within a batch and errors of measurement) to be very small as compared with between-batch variation. In such cases, the *average* of the several observations for a batch may be *treated as an individual observation*. This procedure should, however, be used with great caution; the restrictive conditions just cited should be carefully noted.

The control limits given are three-sigma control limits in all cases.

29. Control Chart for Individuals, X—Using Rational Subgroups

Here the control chart for individuals is commonly used as an adjunct to the more usual $\bar{X}$ and s, or $\bar{X}$ and R, control charts. Proceed exactly as in Sections 9 to 11 (control—no standard given) or Sections 19 to 21 (control—standard given), whichever is applicable, and prepare control charts for $\bar{X}$ and s, or for $\bar{X}$ and R. In addition, prepare a control chart for individuals having the same central line as the $\bar{X}$ chart but compute the control limits as follows:

Formulas for control chart lines.

Chart for Individuals—Associated with Chart for s or R having Sample Size, n

Nature of Data	Central Line	Control Limits: Simplified Formula Using Factors in Table 4	Control Limits: Basic Formula
No Standard Given			
Samples of equal size			
based on s	$\bar{X}$	$\bar{X} \pm E_3\bar{s}$	$\bar{X} \pm 3\bar{s}/c_4$ (34)
based on $\bar{R}$	$\bar{X}$	$\bar{X} \pm E_2\bar{R}$	$\bar{X} \pm 3\bar{R}/d_2$ (35)
Samples of unequal size: σ_e computed from observed values of $\bar{s}$ per Section 9(*b*) or from observed values of $\bar{R}$ per Section 10(*b*)	$\bar{X}$		$\bar{X} \pm 3\sigma_e$ (36)[a]
Standard Given			
Samples of equal or unequal size	$\bar{X}_0'$		$\bar{X}_0' \pm 3\sigma_0'$ (37)

[a] See Example 4, p. 103 for determination of σ_e based on values of s, and Example 6, p. 105 for determination of σ_e based on values of R.

Table 4 gives values of E_2 and E_3 for samples of $n = 10$ or less. More complete values are given in Table 29.

See Examples 22 and 23, pp. 128 and 129.

TABLE 4—*Factors for computing control limits.*[a]

Chart for Individuals—Associated with Chart for s or R Having Sample Size, n

Observations in Samples of Equal Size (from which $\bar{s}$ or $\bar{R}$ has been determined)	2	3	4	5	6	7	8	9	10
Factors for control limits:									
E_3	3.760	3.385	3.256	3.191	3.153	3.129	3.109	3.095	3.084
E_2	2.660	1.772	1.457	1.290	1.184	1.109	1.054	1.010	0.975

[a] Table 4 was revised in 1975 by replacing values of E_1 by values of E_3.

30. Control Chart for Individuals—Using Moving Ranges

(a) No Standard Given

Here the control chart lines are computed from the observed data. In this section the symbol, R, is used to signify the moving range. The control chart lines are as follows.

Formulas for control chart lines.

Chart for Individuals—Using Moving Ranges			
	Central Line	Control Limits	
For individuals	$\bar{X}$	$\bar{X} \pm E_2\bar{R} = \bar{X} \pm 2.66\bar{R}$	(38)
For moving ranges of two observations	$\bar{R}$	$D_4\bar{R}$ and $D_3\bar{R} = 3.27\bar{R}$ and 0	(39)

where

$\bar{X}$ = the average of the individual observations,
$\bar{R}$ = the mean moving range, (see Supplement B, Note 7 for more general discussion) the average of successive differences between the individual observations, and
n = 2 for determining E_2, D_2 and D_4.

See Example 24, p. 130.

(b) Standard Given

When $\bar{X}_0'$ and σ_0' are given, the control chart lines are as follows.

Formulas for control chart lines.

	Central Line	Control Limits	
For individuals	$\bar{X}_0'$	$\bar{X}_0' \pm 3\sigma_0'$	(40)
For moving ranges of two observations	$d_2\sigma_0'$	$D_2\sigma_0'$ and $D_1\sigma_0' = 3.69\sigma_0'$ and 0	(41)

See Example 25, p. 132.

EXAMPLES

31. Illustrative Examples—Control, No Standard Given

Examples 1 to 11, inclusive, illustrate the use of the control chart method of analyzing data for control, when no standard is given (see Sections 7 to 17).

Example 1: Control Charts for $\overline{X}$ *and* s*, Large Samples of Equal Size* (*Section 8*(a))

A manufacturer wished to determine if his product exhibited a state of control. In this case the central lines and control limits were based solely on the data. Table 5 gives observed values of $\overline{X}$ and s for daily samples of n = 50 observations each for ten consecutive days. Figure 2 gives the control charts for $\overline{X}$ and s.

TABLE 5—*Operating characteristic, daily control data.*

Sample	Sample Size, n	Average, $\overline{X}$	Standard Deviation, s
1	50	35.1	5.35
2	50	34.6	4.73
3	50	33.2	3.73
4	50	34.8	4.55
5	50	33.4	4.00
6	50	33.9	4.30
7	50	34.4	4.98
8	50	33.0	5.30
9	50	32.8	3.29
10	50	34.8	3.77
Total	500	340.0	44.00
Average	...	34.0	4.40

Central Lines

For $\overline{X}$: $\overline{\overline{X}} = 34.0$
For s: $\bar{s} = 4.40$

Control Limits
$n = 50$

For $\overline{X}$: $\overline{\overline{X}} \pm 3\dfrac{\bar{s}}{\sqrt{n}} = 34.0 \pm 1.9$,
35.9, and 32.1

For s: $\bar{s} \pm 3\dfrac{\bar{s}}{\sqrt{2(n-1)}} = 4.40 \pm 1.33$,
5.73, and 3.07

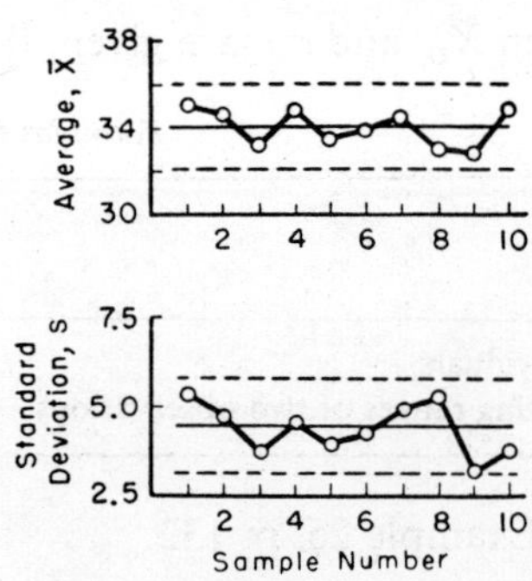

FIG. 2—*Control charts for* $\overline{X}$ *and* s. *Large samples of equal size,* n = *50; no standard given.*

Results—The charts give no evidence of lack of control. Compare with Example 12, in which the same data are used to test product for control at a specified level.

Example 2: Control Charts for $\bar{X}$ *and* s, *Large Samples of Unequal Size (Section 8*(b))

To determine whether there existed any assignable causes of variation in quality for an important operating characteristic of a given product, the inspection results given in Table 6 were obtained from ten shipments whose samples were unequal in size; hence, control limits were computed separately for each sample size. Figure 3 gives the control charts for $\bar{X}$ and s.

TABLE 6—*Operating characteristic, mechanical part.*

Shipment	Sample Size, n	Average, $\bar{X}$	Standard Deviation, s
1	50	55.7	4.35
2	50	54.6	4.03
3	100	52.6	2.43
4	25	55.0	3.56
5	25	53.4	3.10
6	50	55.2	3.30
7	100	53.3	4.18
8	50	52.3	4.30
9	50	53.7	2.09
10	50	54.3	2.67
Total	550	$\Sigma n\bar{X} = 29590.0$	$\Sigma ns = 1864.50$
Weighted average	...	53.8	3.39

Central Lines

For $\bar{X}$: $\bar{\bar{X}} = 53.8$
For s: $\bar{s} = 3.39$

Control Limits

$$\text{For } \bar{X}:\ \bar{\bar{X}} \pm 3\frac{\bar{s}}{\sqrt{n}} = 53.8 \pm \frac{10.17}{\sqrt{n}}$$

$n = 25$: 55.8 and 51.8
$n = 50$: 55.2 and 52.4
$n = 100$: 54.8 and 52.8

$$\text{For } s:\ \bar{s} \pm 3\frac{\bar{s}}{\sqrt{2(n-1)}} = 3.39 \pm \frac{10.17}{\sqrt{2(n-1)}}$$

$n = 25$: 4.86 and 1.92
$n = 50$: 4.42 and 2.36
$n = 100$: 4.11 and 2.67

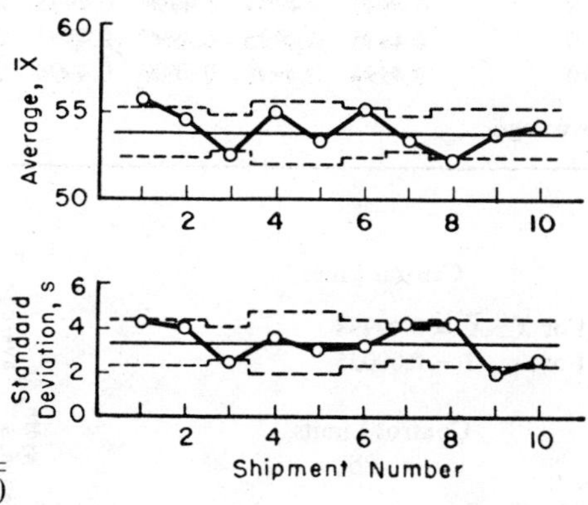

FIG. 3—*Control charts for* $\bar{X}$ *and* s. *Large samples of unequal size,* n = *25, 50, 100; no standard given.*

Results—Lack of control is indicated with respect to both $\bar{X}$ and s. Corrective action is needed to reduce the variability between shipments.

Example 3: Control Charts for $\bar{X}$ *and* s, *Small Samples of Equal Size* (*Section* 9(a))

Table 7 gives the width in inches to the nearest 0.0001 in. measured prior to exposure for ten sets of corrosion specimens of Grade BB zinc. These two groups of five sets each were selected for illustrative purposes from a large number of sets of specimens consisting of six specimens each used in atmosphere exposure tests sponsored by ASTM. In each of the two groups the five sets correspond to five different millings that were employed in the preparation of the specimens. Figure 4 shows control charts for $\bar{X}$ and s.

TABLE 7—*Width in inches, specimens of Grade BB zinc.*

Set	Measured Values						Average, $\bar{X}$	Standard Deviation, s	Range, R
	X_1	X_2	X_3	X_4	X_5	X_6			
Group 1									
1	0.5005	0.5000	0.5008	0.5000	0.5005	0.5000	0.50030	0.00035	0.0008
2	0.4998	0.4997	0.4998	0.4994	0.4999	0.4998	0.49973	0.00018	0.0005
3	0.4995	0.4995	0.4995	0.4995	0.4995	0.4996	0.49952	0.00004	0.0001
4	0.4998	0.5005	0.5005	0.5002	0.5003	0.5004	0.50028	0.00026	0.0007
5	0.5000	0.5005	0.5008	0.5007	0.5008	0.5010	0.50063	0.00035	0.0010
Group 2									
6	0.5008	0.5009	0.5010	0.5005	0.5006	0.5009	0.50078	0.00019	0.0005
7	0.5000	0.5001	0.5002	0.4995	0.4996	0.4997	0.49985	0.00029	0.0007
8	0.4993	0.4994	0.4999	0.4996	0.4996	0.4997	0.49958	0.00021	0.0006
9	0.4995	0.4995	0.4997	0.4992	0.4995	0.4992	0.49943	0.00020	0.0005
10	0.4994	0.4998	0.5000	0.4990	0.5000	0.5000	0.49970	0.00041	0.0010
Average							0.49998	0.00025	0.00064

Central Lines

For $\bar{X}$: $\bar{\bar{X}} = 0.49998$
For s: $\bar{s} = 0.00025$

Control Limits
$n = 6$

For $\bar{X}$: $\bar{\bar{X}} \pm A_3\bar{s} =$
$0.49998 \pm (1.287)(0.00025)$
0.50030 and 0.49966

For s: $B_4\bar{s}$ and $B_3\bar{s} =$
(1.970) (0.00025) and
(0.030) (0.00025)
0.00049 and 0.00001

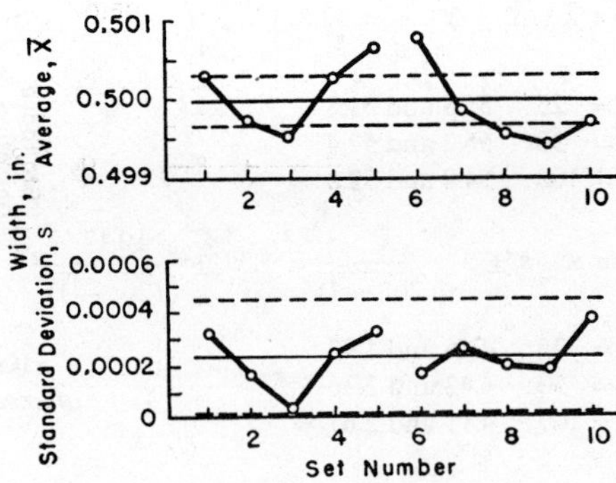

FIG. 4—*Control chart for* $\bar{X}$ *and* s. *Small samples of equal size,* n = 6; *no standard given.*

Results—The chart for averages indicates the presence of assignable causes of variation in width, $\overline{X}$, from set to set, that is, from milling to milling. The pattern of points for averages indicates a systematic pattern of width values for the five millings, a factor that required recognition in the analysis of the corrosion test results.

Example 4: Control Charts for $\overline{\text{X}}$ *and* s, *Small Samples of Unequal Size (Section* 9(b))

Table 8 gives interlaboratory calibration check data on 21 horizontal tension testing machines. The data represent tests on No. 16 wire. The procedure is similar to that given in Example 3, but indicates a suggested method of computation when the samples are not equal in size. Figure 5 gives control charts for $\overline{X}$ and s.

$$\sigma_e = \frac{1}{21}\left(\frac{2.41}{0.9213} + \frac{15.34}{0.9400}\right) = 0.902$$

TABLE 8—*Interlaboratory calibration, horizontal tension testing machines.*

Machine	Number of Tests	Test Value 1	2	3	4	5	Average, $\overline{X}$	Standard Deviation, s $n=4$	$n=5$	Range, R $n=4$	$n=5$
1	5	73	73	73	75	75	73.8	...	1.10	...	2
2	5	70	71	71	71	72	71.0	...	0.71	...	2
3	5	74	74	74	74	75	74.2	...	0.45	...	1
4	5	70	70	70	72	73	71.0	...	1.41	...	3
5	5	70	70	70	70	70	70.0	...	0	...	0
6	5	65	65	66	69	70	67.0	...	2.35	...	5
7	4	72	72	74	76	...	73.5	1.91	...	4	...
8	5	69	70	71	73	73	71.2	...	1.79	...	4
9	5	71	71	71	71	72	71.2	...	0.45	...	1
10	5	71	71	71	71	72	71.2	...	0.45	...	1
11	5	71	71	72	72	72	71.6	...	0.55	...	1
12	5	70	71	71	72	72	71.2	...	0.55	...	2
13	5	73	74	74	75	75	74.2	...	0.84	...	2
14	5	74	74	75	75	75	74.6	...	0.55	...	1
15	5	72	72	72	73	73	72.4	...	0.55	...	1
16	4	75	75	75	76	...	75.3	0.50	...	1	...
17	5	68	69	69	69	70	69.0	...	0.71	...	2
18	5	71	71	72	72	73	71.8	...	0.84	...	2
19	5	72	73	73	73	73	72.8	...	0.45	...	1
20	5	68	69	70	71	71	69.8	...	1.30	...	3
21	5	69	69	69	69	69	69.0	...	0	...	0
Total	103				weighted average $\overline{X}$ =		71.65	2.41	15.34	5	34

Central Lines

For $\overline{X}$: $\overline{\overline{X}} = 71.65$

For s: $n = 4$: $\bar{s} = c_4\sigma_e = (0.9213)(0.902) = 0.831$

$n = 5$: $\bar{s} = c_4\sigma_e = (0.9400)(0.902) = 0.848$

Control Limits

For $\overline{X}$: $n = 4$: $\overline{\overline{X}} \pm A_3\bar{s} =$ $71.65 \pm (1.628)(0.831)$, 73.0, and 70.3

$n = 5$: $\overline{\overline{X}} \pm A_3\bar{s} =$ $71.65 \pm (1.427)(0.848)$, 72.9, and 70.4

For s: $n = 4$: $B_4\bar{s}$ and $B_3\bar{s} =$ (2.266)(0.831) and (0)(0.831), 1.88, and 0

$n = 5$: $B_4\bar{s}$ and $B_3\bar{s} =$ (2.089)(0.848) and (0)(0.848), 1.77, and 0

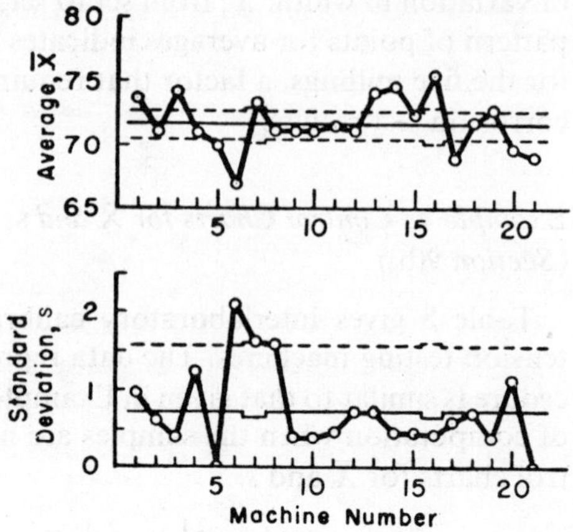

FIG. 5—*Control charts for* $\overline{X}$ *and* s. *Small samples of unequal size,* n = *4, 5; no standard given.*

Results—The calibration levels of machines were not controlled at a common level; the averages of six machines are above and the averages of five machines are below control limits. Likewise, there is an indication that the variability *within* machines is not in statistical control, since three machines, Numbers 6, 7, and 8, have standard deviations outside control limits.

Example 5: Control Charts for $\overline{X}$ *and* R, *Small Samples of Equal Size (Section 10*(a))

Same data as in Example 3, Table 7. Use is made of control charts for averages and ranges rather than for averages and standard deviations. Figure 6 shows control charts for $\overline{X}$ and R.

Central Lines

For $\overline{X}$: $\overline{\overline{X}} = 0.49998$

For R: $\overline{R} = 0.00064$

Control Limits

$n = 6$

For $\overline{X}$: $\overline{\overline{X}} \pm A_2\overline{R} =$ $0.49998 \pm (0.483)(0.00064)$, 0.50029, and 0.49967

For R: $D_4\overline{R}$ and $D_3\overline{R} =$ (2.004)(0.00064) and (0)(0.00064), 0.00128, and 0

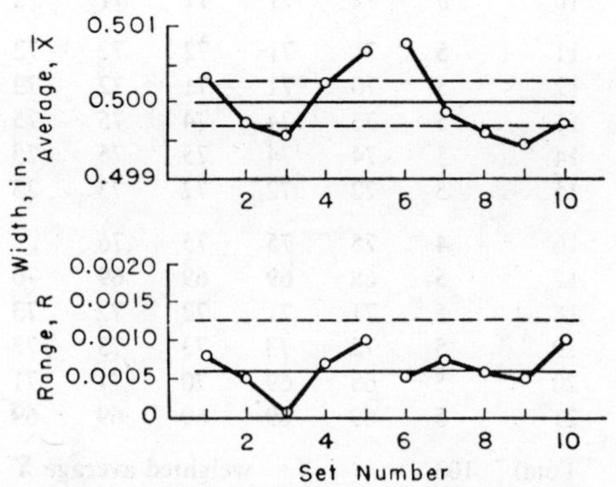

FIG. 6—*Control charts for* $\overline{X}$ *and* R. *Small samples of equal size,* n = *6; no standard given.*

Results—The results are practically identical in all respects with those obtained by using averages and standard deviations, Fig. 4, Example 3.

Example 6: Control Charts for X̄ *and* R, *Small Samples of Unequal Size* (*Section 10*(b))

Same data as in Example 4, Table 8. In the analysis and control charts, the range is used instead of the standard deviation. The procedure is similar to that given in Example 5, but indicates a suggested method of computation when samples are not equal in size. Figure 7 gives control charts for $\bar{X}$ and R.

σ_e is determined from the tabulated ranges given in Example 4, using a similar procedure to that given in Example 4 for standard deviations where samples are not equal in size, that is

$$\sigma_e = \frac{1}{21}\left(\frac{5}{2.059} + \frac{34}{2.326}\right) = 0.812$$

Central Lines

For $\bar{X}$: $\bar{\bar{X}} = 71.65$

For R: $n = 4$: $\bar{R} = d_2\sigma_e = (2.059)(0.812) = 1.67$

$n = 5$: $\bar{R} = d_2\sigma_e = (2.326)(0.812) = 1.89$

Control Limits

For $\bar{X}$: $n = 4$: $\bar{\bar{X}} \pm A_2\bar{R} = 71.65 \pm (0.729)(1.67)$, 72.9, and 70.4

$n = 5$: $\bar{\bar{X}} \pm A_2\bar{R} = 71.65 \pm (0.577)(1.89)$, 72.7, and 70.6

For R: $n = 4$: $D_4\bar{R}$ and $D_3\bar{R}$ = (2.282)(1.67) and (0)(1.67), 3.8, and 0

$n = 5$: $D_4\bar{R}$ and $D_3\bar{R}$ = (2.114)(1.89) and (0)(1.89), 4.0, and 0

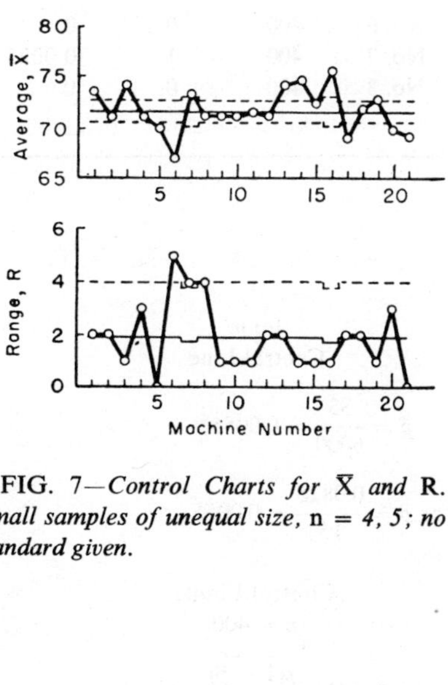

FIG. 7—*Control Charts for* X̄ *and* R. *Small samples of unequal size,* n = *4, 5; no standard given.*

Results—The results are practically identical in all respects with those obtained by using averages and standard deviations, Fig. 5, Example 4.

Example 7: Control Charts for p, *Samples of Equal Size* (*Section 13*(a)) *and* pn, *Samples of Equal Size* (*Section 14*)

Table 9 gives the number of defectives found in inspecting a series of 15 consecutive lots of galvanized washers for finish defects such as exposed steel, rough galvanizing. The lots were of almost the same size and a constant sample size, $n = 400$ was used. The fraction defective for each sample was determined by dividing the number of defectives found, pn, by the sample size, n, and is listed in the table. Figure 8 gives the control chart for p, and Fig. 9 gives the control chart for pn. Note that these two charts are identical except for the vertical scale.

TABLE 9—*Finish defects, galvanized washers.*

Lot	Sample Size, n	Number of Defectives, pn	Fraction Defective, p	Lot	Sample Size, n	Number of Defectives, pn	Fraction Defective, p
No. 1	400	1	0.0025	No. 9	400	8	0.0200
No. 2	400	3	0.0075	No. 10	400	5	0.0125
No. 3	400	0	0				
No. 4	400	7	0.0175	No. 11	400	2	0.0050
No. 5	400	2	0.0050	No. 12	400	0	0
				No. 13	400	1	0.0025
No. 6	400	0	0	No. 14	400	0	0
No. 7	400	1	0.0025	No. 15	400	3	0.0075
No. 8	400	0	0				
				Total	6000	33	0.0825

(*a*) p
Central Line

$$\bar{p} = \frac{33}{6000} = 0.0055$$

$$\text{or } \bar{p} = \frac{0.0825}{15} = 0.0055$$

Control Limits
$n = 400$

$$\bar{p} \pm 3\sqrt{\frac{\bar{p}(1-\bar{p})}{n}} =$$

$$0.0055 \pm 3\sqrt{\frac{0.0055\,(0.9945)}{400}} =$$

0.0055 ± 0.0111, 0.0166, and 0

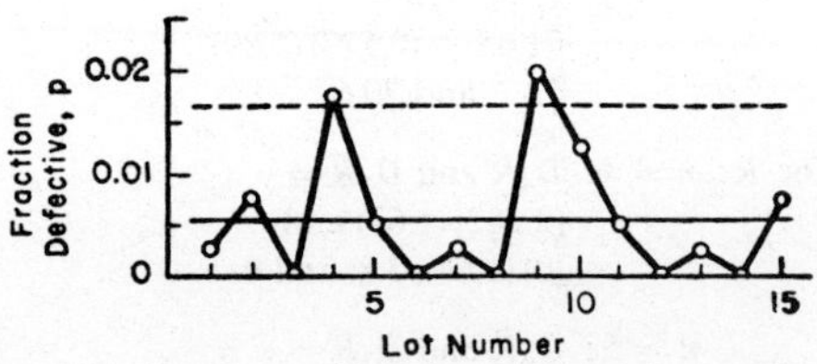

FIG. 8—*Control chart for* p. *Samples of equal size,* n = *400; no standard given.*

(*b*) *pn*
Central Line
$n = 400$

$$\bar{p}n = \frac{33}{15} = 2.2$$

Control Limits
$n = 400$

$$\bar{p}n \pm 3\sqrt{\bar{p}n} = 2.2 \pm 4.4, 6.6, \text{ and } 0$$

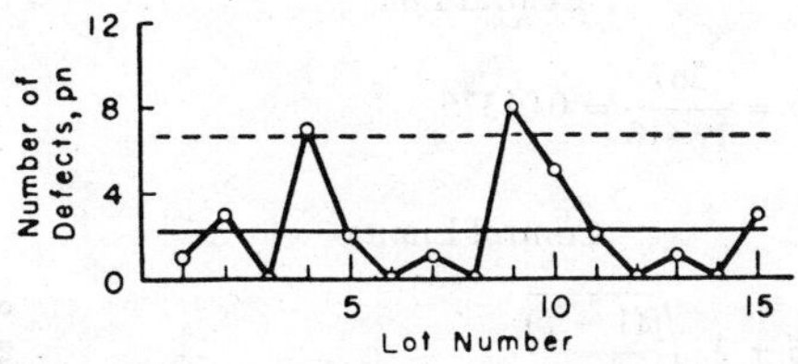

FIG. 9—*Control chart for* pn. *Samples of equal size,* n = *400; no standard given.*

Results—Lack of control is indicated; points for lots Numbers 4 and 9 are outside control limits.

Example 8: Control Chart for p, *Samples of Unequal Size* (*Section 13*(b))

Table 10 gives inspection results for surface defects on 31 lots of a certain type of galvanized hardware. The lot sizes varied considerably and corresponding variations in sample sizes were used. Figure 10 gives the control chart for fraction defective, p. In practice, results are commonly expressed in "percent defective," using scale values of 100 times p.

TABLE 10—*Surface defects, galvanized hardware.*

Lot	Sample Size, n	Number of Defectives, pn	Fraction Defective, p	Lot	Sample Size, n	Number of Defectives, pn	Fraction Defective, p
No. 1	580	9	0.0155	No. 16	330	4	0.0121
No. 2	550	7	0.0127	No. 17	330	2	0.0061
No. 3	580	3	0.0052	No. 18	640	4	0.0063
No. 4	640	9	0.0141	No. 19	580	7	0.0121
No. 5	880	13	0.0148	No. 20	550	9	0.0164
No. 6	880	14	0.0159	No. 21	510	7	0.0137
No. 7	640	14	0.0219	No. 22	640	12	0.0188
No. 8	550	10	0.0182	No. 23	200	7	0.0350
No. 9	580	12	0.0207	No. 24	330	5	0.0152
No. 10	880	14	0.0159	No. 25	880	18	0.0205
No. 11	800	6	0.0075	No. 26	880	7	0.0080
No. 12	800	12	0.0150	No. 27	800	8	0.0100
No. 13	580	7	0.0121	No. 28	580	8	0.0138
No. 14	580	11	0.0190	No. 29	880	15	0.0170
No. 15	550	5	0.0091	No. 30	880	3	0.0034
				No. 31	330	5	0.0152
				Total	19 410	267	

Central Line

$$\bar{p} = \frac{267}{19\,410} = 0.01376$$

Control Limits

$$\bar{p} \pm 3\sqrt{\frac{\bar{p}(1-\bar{p})}{n}}$$

For $n = 200$

$$0.01376 \pm 3\sqrt{\frac{0.01376\,(0.98624)}{200}} =$$

$$0.01376 \pm 3(0.008237) =$$
$$0.01376 \pm 0.02471, 0.03847, \text{ and } 0$$

For $n = 880$

$$0.01376 \pm 3\sqrt{\frac{0.01376\,(0.98624)}{880}} =$$

$$0.01376 \pm 3(0.003927) =$$
$$0.01376 \pm 0.01178,$$
$$0.02554, \text{ and } 0.00198$$

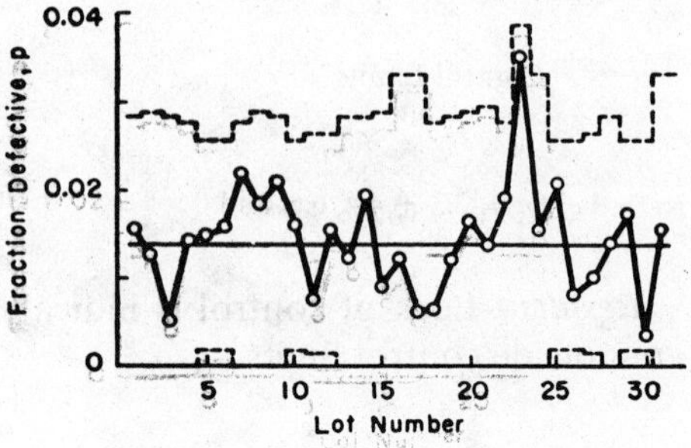

FIG. 10—*Control chart for* p. *Samples of unequal size,* n = *200 to 880; no standard given.*

Results—A state of control may be assumed to exist since 25 consecutive subgroups fall within 3-sigma control limits.

TABLE 11—*Number of defects in consecutive samples of ten units each—burlap bags.*

Sample	Total Defects in Sample, c	Defects per Unit, u	Sample	Total Defects in Sample, c	Defects per Unit, u
1	17	1.7	13	8	0.8
2	14	1.4	14	11	1.1
3	6	0.6	15	18	1.8
4	23	2.3	16	13	1.3
5	5	0.5	17	22	2.2
6	7	0.7	18	6	0.6
7	10	1.0	19	23	2.3
8	19	1.9	20	22	2.2
9	29	2.9	21	9	0.9
10	18	1.8	22	15	1.5
11	25	2.5	23	20	2.0
12	5	0.5	24	6	0.6
			25	24	2.4
			Total	375	37.5

Example 9: Control Charts for u, *Samples of Equal Size* (*Section 15*(a)), *and* c, *Samples of Equal Size* (*Section 16*(a))

Table 11 gives inspection results in terms of defects observed in the inspection of 25 consecutive lots of burlap bags. Since the number of bags in each lot differed slightly, a constant sample size, $n = 10$ was used. All defects were counted even though 2 or more defects of the same or different kinds occurred on the same bag. The defects per unit value for each sample was determined by dividing the number of defects found by the sample size and is listed in the table. Figure 11 gives the control chart for u, and Fig. 12 gives the control chart for c. Note that these two charts are identical except for the vertical scale.

(*a*) u

Central Line

$$\bar{u} = \frac{37.5}{25} = 1.5$$

Control Limits

$n = 10$

$$\bar{u} \pm 3\sqrt{\frac{\bar{u}}{n}} =$$

$$1.50 \pm 3\sqrt{0.150} =$$
$$1.50 \pm 1.16, 2.66, \text{ and } 0.34$$

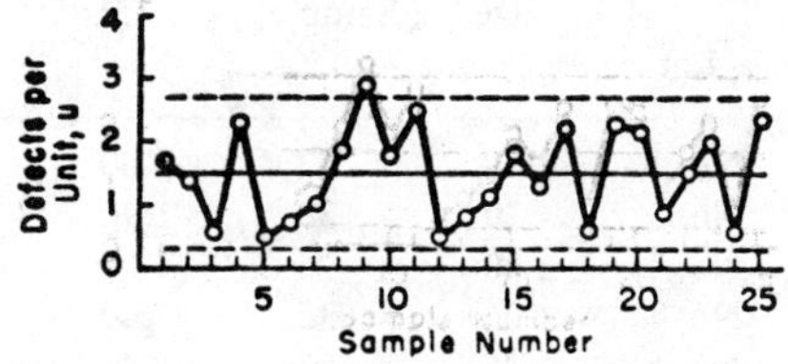

FIG. 11—*Control chart for* u. *Samples of equal size,* n = *10; no standard given.*

(*b*) c

Central Line

$$\bar{c} = \frac{375}{25} = 15.0$$

Control Limits

$n = 10$

$$\bar{c} \pm 3\sqrt{\bar{c}} =$$

$$15.0 \pm 3\sqrt{15} =$$
$$15.0 \pm 11.6, 26.6, \text{ and } 3.4$$

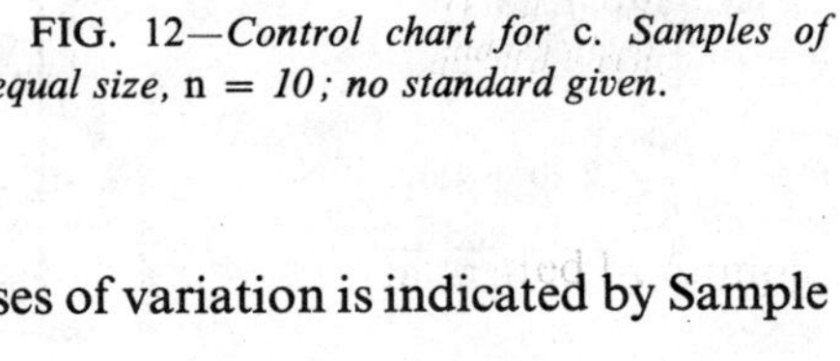

FIG. 12—*Control chart for* c. *Samples of equal size,* n = *10; no standard given.*

Results—Presence of assignable causes of variation is indicated by Sample 9.

Example 10: Control Chart for u, *Samples of Unequal Size* (*Section 15*(b))

Table 12 gives inspection results for 20 lots of different sizes for which 3 different sample sizes were used, 20, 25, and 40. The observed defects in this inspection cover all of the specified characteristics of a complex machine

(Type A), including a large number of dimensional, operational, as well as physical and finish requirements. Because of the large number of tests and measurements required as well as possible occurrences of minor observed irregularities, the expectancy of defects per unit is high, although the majority of such defects are of minor seriousness. The defects per unit value for each sample, number of defects in sample divided by number of units in sample, was determined and these values are listed in the last column of the table. Figure 13 gives the control chart for u with control limits corresponding to the three different sample sizes.

TABLE 12—*Number of defects in samples from 20 successive lots of Type A machines.*

Lot	Sample Size, n	Total Defects in Sample, c	Defects per Unit, u	Lot	Sample Size, n	Total Defects in Sample, c	Defects per Unit, u
No. 1	20	72	3.60	No. 11	25	47	1.88
No. 2	20	38	1.90	No. 12	25	55	2.20
No. 3	40	76	1.90	No. 13	25	49	1.96
No. 4	25	35	1.40	No. 14	25	62	2.48
No. 5	25	62	2.48	No. 15	25	71	2.84
No. 6	25	81	3.24	No. 16	20	47	2.35
No. 7	40	97	2.42	No. 17	20	41	2.05
No. 8	40	78	1.95	No. 18	20	52	2.60
No. 9	40	103	2.58	No. 19	40	128	3.20
No. 10	40	56	1.40	No. 20	40	84	2.10
				Total	580	1334	

Central Line

$$\bar{u} = \frac{1334}{580} = 2.30$$

Control Limits

$n = 20$

$$\bar{u} \pm 3\sqrt{\frac{\bar{u}}{n}} = 2.30 \pm 1.02,$$
3.32, and 1.28

$n = 25$

$$\bar{u} \pm 3\sqrt{\frac{\bar{u}}{n}} = 2.30 \pm 0.91,$$
3.21, and 1.39

$n = 40$

$$\bar{u} \pm 3\sqrt{\frac{\bar{u}}{n}} = 2.30 \pm 0.72,$$
3.02, and 1.58

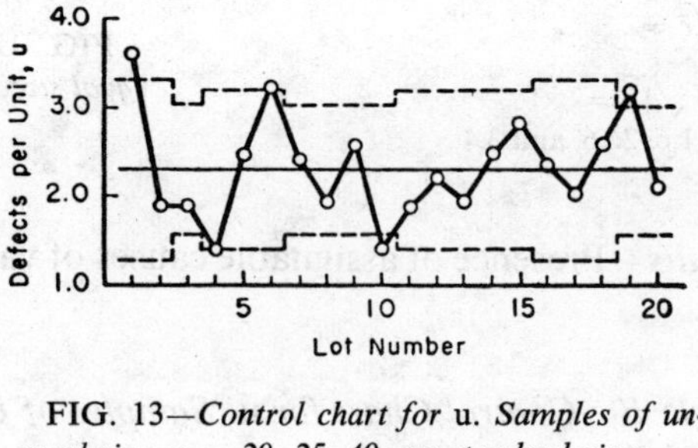

FIG. 13—*Control chart for* u. *Samples of unequal size,* n = *20, 25, 40; no standard given.*

Results—Lack of control of quality is indicated; plotted points for lots Numbers 1, 6, and 19 are above the upper control limit and the point for lot Number 10 is below the lower control limit.

Example 11: Control Charts for c, *Samples of Equal Size* (*Section 16*(a))

Table 13 gives the results of continuous testing of a certain type of rubber-covered wire at specified test voltage. This test causes breakdowns at weak spots in the insulation which are cut out before shipment of wire in short coil lengths. The original data obtained consisted of records of the number of breakdowns in successive lengths of 1000 ft each. There may be 0, 1, 2, 3, . . . , etc. breakdowns per length, depending on the number of weak spots in the insulation. Such data might also have been tabulated as number of breakdowns in successive lengths of 100 ft each, 500 ft each, etc. Here there is no natural unit of product (such as 1 in., 1 ft, 10 ft, 100 ft, etc.), in respect

TABLE 13—*Number of breakdowns in successive lengths of 5 000 ft each and 10 000 ft each for rubber-covered wire.*

Length No.	Number of Break-downs	Length No.	Number of Break-downs	Length No.	Number of Break-downs	Length No.	Number of Break-downs	Length No.	Number of Break-downs
				Lengths of 5 000 ft Each					
1	0	13	1	25	0	37	5	49	5
2	1	14	1	26	0	38	7	50	4
3	1	15	2	27	9	39	1	51	2
4	0	16	4	28	10	40	3	52	0
5	2	17	0	29	8	41	3	53	1
6	1	18	1	30	8	42	2	54	2
7	3	19	1	31	6	43	0	55	5
8	4	20	0	32	14	44	1	56	9
9	5	21	6	33	0	45	5	57	4
10	3	22	4	34	1	46	3	58	2
11	0	23	3	35	2	47	4	59	5
12	1	24	2	36	4	48	3	60	3
Total								60	187
				Lengths of 10 000 ft Each					
1	1	7	2	13	0	19	12	25	9
2	1	8	6	14	19	20	4	26	2
3	3	9	1	15	16	21	5	27	3
4	7	10	1	16	20	22	1	28	14
5	8	11	10	17	1	23	8	29	6
6	1	12	5	18	6	24	7	30	8
Total								30	187

to the quality characteristic "breakdown" since failures may occur at any point. Since the original data were given in terms of 1000-ft lengths, a control chart might have been maintained for "number of breakdowns in successive lengths of 1000 ft each." So many points were obtained during a short period of production by using the 1000-ft length as a unit and the expectancy in term of number of breakdowns per length was so small that longer unit lengths were tried. Table 13 gives (*a*) the "number of breakdowns in successive lengths of 5000 ft each," and (*b*) the "number of breakdowns in successive lengths of 10 000 ft each." Figure 14 shows the control chart for *c* where the unit selected is 5000 ft, and Fig. 15 shows the control chart for *c* where the unit selected is 10 000 ft. The standard unit length finally adopted for control purposes was 10 000 ft for "breakdown."

(*a*) Lengths of 5000 ft Each

Central Line

$$c = \frac{187}{60} = 3.12$$

Control Limits

$$\bar{c} \pm 3\sqrt{\bar{c}} =$$

$$3.12 \pm 3\sqrt{3.12},$$

8.42, and 0

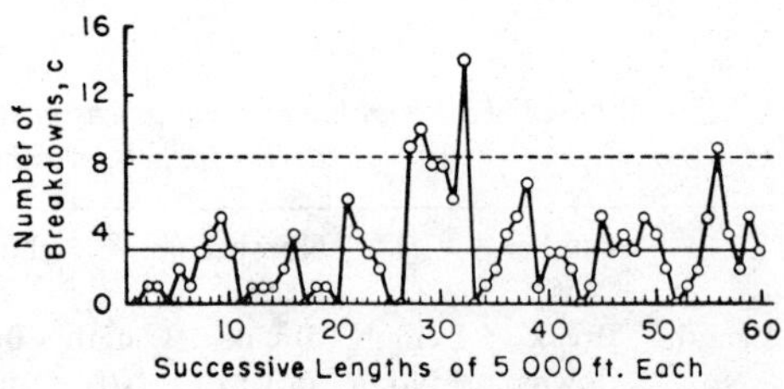

FIG. 14—*Control chart for* c. *Samples of equal size,* n = *1 standard length of 5000 ft; no standard given.*

Results—Presence of assignable causes of variation is indicated by lengths Numbers 27, 28, 32, and 56 falling above upper control limit.

(*b*) Lengths of 10 000 ft Each

Central Line

$$\bar{c} = \frac{187}{30} = 6.23$$

Control Limits

$$\bar{c} \pm 3\sqrt{\bar{c}} =$$

$$6.23 \pm 3\sqrt{6.23},$$

13.72, and 0

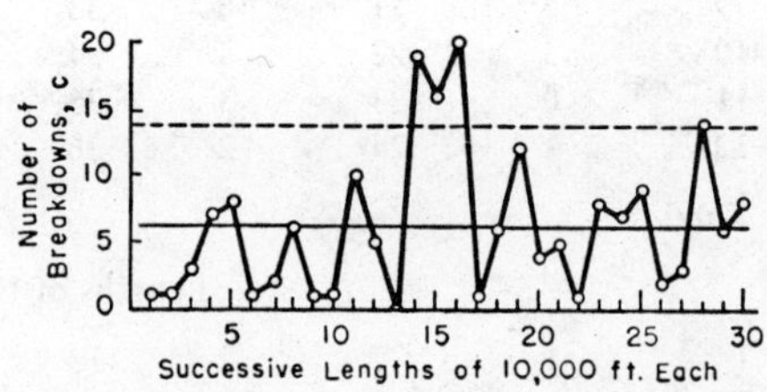

FIG. 15—*Control chart for* c. *Samples of equal size,* n = *1 standard length of 10 000 ft; no standard given.*

Results—Presence of assignable causes of variation is indicated by lengths Numbers 14, 15, 16, and 28 falling above upper control limit.

32 Illustrative Examples—Control With Respect to a Given Standard

Examples 12 to 21, inclusive, illustrate the use of the control chart method of analyzing data for control with respect to a given standard (see Sections 18 to 27).

Example 12: Control Charts for $\bar{X}$ *and* s*, Large Samples of Equal Size (Section 19)*

A manufacturer attempted to maintain an aimed-at distribution of quality for a certain operating characteristic. The objective standard distribution which served as a target was defined by standard values: $\bar{X}_0' = 35.00$ lb, and $\sigma_0' = 4.20$ lb. Table 14 gives observed values of $\bar{X}$ and s for daily samples of $n = 50$ observations each for ten consecutive days. These data are the same as used in Example 1 and presented as Table 5. Figure 16 gives control charts for $\bar{X}$ and s.

TABLE 14—*Operating characteristic, daily control data.*

Sample	Sample Size, n	Average, $\bar{X}$	Standard Deviation, s
1	50	35.1	5.35
2	50	34.6	4.73
3	50	33.2	3.73
4	50	34.8	4.55
5	50	33.4	4.00
6	50	33.9	4.30
7	50	34.4	4.98
8	50	33.0	5.30
9	50	32.8	3.29
10	50	34.8	3.77

Central Lines

For $\bar{X}$: $\bar{X}_0' = 35.00$
For s: $\sigma_0' = 4.20$

Control Limits
$n = 50$

For $\bar{X}$: $\bar{X}_0' \pm 3\dfrac{\sigma_0'}{\sqrt{n}} = 35.00 \pm 1.8$,
36.8, and 33.2

For s: $\sigma_0' \pm 3\dfrac{\sigma_0'}{\sqrt{2(n-1)}} = 4.20 \pm 1.27$,
5.47, and 2.93

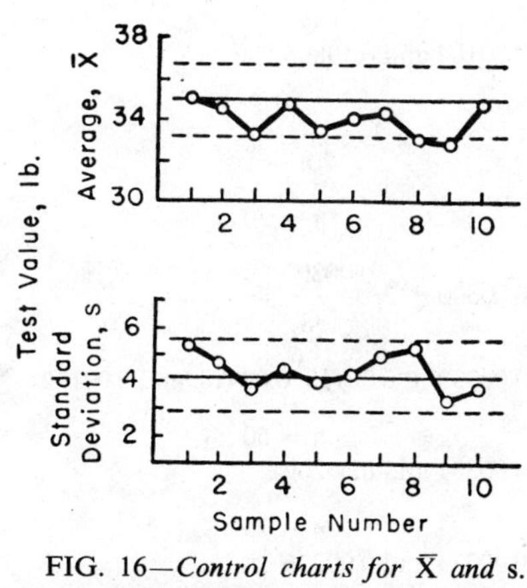

FIG. 16—*Control charts for* $\bar{X}$ *and* s*. Large samples of equal size,* n = *50:* $\bar{X}_0'$*,* σ_0' *given.*

Results—Lack of control at standard level is indicated on the eighth and ninth days. Compare with Example 1 in which the same data were analyzed for control without specifying a standard level of quality.

Example 13: Control Charts for $\bar{X}$ *and* s, *Large Samples of Unequal Size (Section 19)*

For a product it was desired to control a certain critical dimension, the diameter, with respect to day to day variation. Daily sample sizes of 30, 50, or 75 were selected and measured, the number taken depending on the quantity produced per day. The desired level was $\bar{X}_0' = 0.20000$ in. with $\sigma_0' = 0.00300$ in. Table 15 gives observed values of $\bar{X}$ and s for the samples from 10 successive days' production. Figure 17 gives the control charts for $\bar{X}$ and s.

Central Lines

For $\bar{X}$: $\bar{X}_0' = 0.20000$
For s: $\bar{s} = \sigma_0' = 0.00300$

Control Limits

For $\bar{X}$: $\bar{X}_0' \pm 3 \dfrac{\sigma_0'}{\sqrt{n}}$

$n = 30$

$$0.20000 \pm 3 \frac{0.00300}{\sqrt{30}} =$$

0.20000 ± 0.00164, 0.20164, and 0.19836

$n = 50$

0.20127 and 0.19873

$n = 75$

0.20104 and 0.19896

For s: $\bar{s} \pm 3 \dfrac{\sigma_0'}{\sqrt{2(n-1)}}$

$n = 30$

$$0.00300 \pm 3 \frac{0.00300}{\sqrt{58}} =$$

0.00300 ± 0.00118, 0.00418, and 0.00182

$n = 50$

0.00391 and 0.00209

$n = 75$

0.00374 and 0.00226

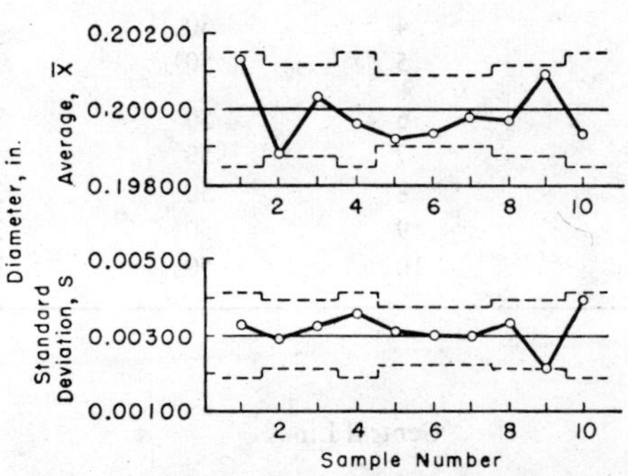

FIG. 17—*Control charts for* $\bar{X}$ *and* s. *Large samples of unequal size,* n = *30, 50, 75;* X_0', σ_0' *given.*

TABLE 15—*Diameter in inches, control data.*

Sample	Sample Size n	Average, $\overline{X}$	Standard Deviation, s
1	30	0.20133	0.00330
2	50	0.19886	0.00292
3	50	0.20037	0.00326
4	30	0.19965	0.00358
5	75	0.19923	0.00313
6	75	0.19934	0.00306
7	75	0.19984	0.00299
8	50	0.19974	0.00335
9	50	0.20095	0.00221
10	30	0.19937	0.00397

Results—The charts give no evidence of significant deviations from standard values.

Example 14: Control Chart for $\overline{X}$ *and* s, *Small Samples of Equal Size* (*Section 19*)

Same product and characteristic as in Example 13, but in this case it is desired to control the diameter of this product with respect to sample variations during each day, since samples of 10 were taken at definite intervals each day. The desired level is $\overline{X}_0' = 0.20000$ in. with $\sigma_0' = 0.00300$ in. Table 16 gives observed values of $\overline{X}$ and s for 10 samples of 10 each taken during a single day. Figure 18 gives the control charts for $\overline{X}$ and s.

TABLE 16—*Control data for one day's product.*

Sample	Sample Size, n	Average, $\overline{X}$	Standard Deviation, s
1	10	0.19838	0.00350
2	10	0.20126	0.00304
3	10	0.19868	0.00333
4	10	0.20071	0.00337
5	10	0.20050	0.00159
6	10	0.20137	0.00104
7	10	0.19883	0.00299
8	10	0.20218	0.00327
9	10	0.19868	0.00431
10	10	0.19968	0.00356

Central Lines

For $\overline{X}$: $\overline{X}_0' = 0.20000$

$n = 10$

For s: $c_4\sigma_0' = (0.9727)(0.00300) = 0.00292$

Control Limits

$n = 10$

For $\overline{X}$: $\overline{X}_0' \pm A\sigma_0' =$
$0.20000 \pm (0.949)(0.00300)$,
0.20285, and 0.19715

For s: $B_6\sigma_0'$ and $B_5\sigma_0' =$
(1.669) (0.00300) and
(0.276) (0.00300),
0.00501, and 0.00083

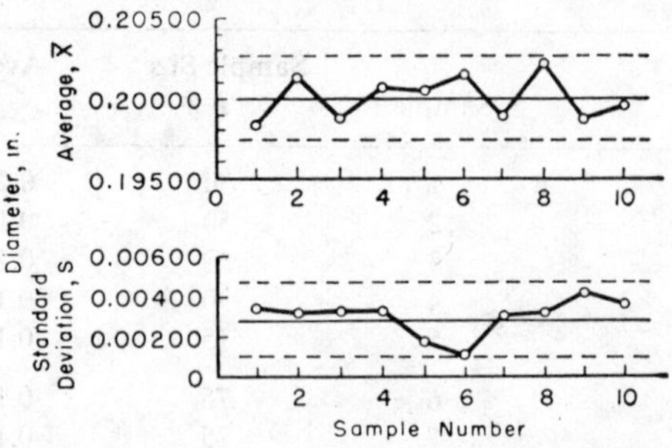

FIG. 18—*Control charts for* $\overline{X}$ *and* s. *Small samples of equal size,* $n = 10$; $\overline{X}_0'$, σ_0' *given.*

Results—No lack of control is indicated.

Example 15: Control Chart for $\overline{X}$ *and* s, *Small Samples of Unequal Size (Section 19)*

A manufacturer wished to control the resistance of a certain product after it had been operating for 100 h, where $\overline{X}_0' = 150\ \Omega$ and $\sigma_0' = 7.5\ \Omega$. From each of 15 consecutive lots, he selected a random sample of 5 units and subjected them to the operating test for 100 h. Due to mechanical failures, some of the units in the sample failed before the completion of 100 h of operation. Table 17 gives the averages and standard deviations for the 15 samples together with their sample sizes. Figure 19 gives the control charts for $\overline{X}$ and s.

TABLE 17—*Resistance in ohms after 100-hr operation, lot by lot control data.*

Sample	Sample Size, n	Average, $\overline{X}$	Standard Deviation, s	Sample	Sample Size, n	Average, $\overline{X}$	Standard Deviation, s
1	5	154.6	12.20	9	5	156.2	8.92
2	5	143.4	9.75	10	4	137.5	3.24
3	4	160.8	11.20	11	5	153.8	6.85
4	3	152.7	7.43	12	5	143.4	7.64
5	5	136.0	4.32	13	4	156.0	10.18
6	3	147.3	8.65	14	5	149.8	8.86
7	3	161.7	9.23	15	3	138.2	7.38
8	5	151.0	7.24				

Central Lines

For $\bar{X}$: $\bar{X}_0' = 150$

For s:

$$n = 3$$
$$\bar{s} = c_4\sigma_0' = (0.8862)(7.5) = 6.65$$

$$n = 4$$
$$\bar{s} = c_4\sigma_0' = (0.9213)(7.5) = 6.91$$

$$n = 5$$
$$\bar{s} = c_4\sigma_0' = (0.9400)(7.5) = 7.05$$

Control Limits

For $\bar{X}$:

$$n = 3$$
$$\bar{X}_0' \pm A\sigma_0' = 150 \pm 1.732\,(7.5),$$
163.0, and 137.0

$$n = 4$$
$$\bar{X}_0' \pm A\sigma_0' = 150 \pm 1.500\,(7.5),$$
161.2, 138.8

$$n = 5$$
$$\bar{X}_0' \pm A\sigma_0' = 150 \pm 1.342\,(7.5),$$
160.1, and 139.9

For s:

$n = 3$: $B_6\sigma_0'$ and $B_5\sigma_0' = (2.276)(7.5)$ and $(0)(7.5)$, 17.07, and 0

$n = 4$: $B_6\sigma_0'$ and $B_5\sigma_0' = (2.088)(7.5)$ and $(0)(7.5)$, 15.66, and 0

$n = 5$: $B_6\sigma_0'$ and $B_5\sigma_0' = (1.964)(7.5)$ and $(0)(7.5)$, 14.73, and 0

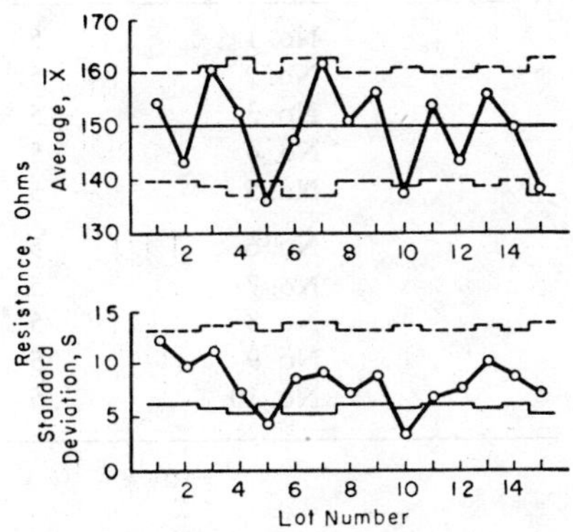

FIG. 19—*Control charts for* $\bar{X}$ *and* s. *Small samples of unequal size,* n = *3, 4, 5;* X_0', σ_0' *given.*

Results—Evidence of lack of control is indicated since samples from lots Numbers 5 and 10 have averages below their lower control limit. No standard deviation values are outside their control limits. Corrective action is required to reduce the variation between lot averages.

Example 16: Control Charts for $\bar{X}$ *and* R, *Small Samples of Equal Size (Section 20)*

Consider the same problem as in Example 12 where $\bar{X}_0' = 35.00$ lb and $\sigma_0' = 4.20$ lb. The manufacturer wished to control variations in quality from lot to lot by taking a small sample from each lot. Table 18 gives observed values of $\bar{X}$ and R for samples of $n = 5$ each, selected from ten consecutive lots. Since the sample size n is less than 10, actually 5, he elected to use control charts for $\overline{X}$ and R rather than for $\overline{X}$ and s. Figure 20 gives the control charts for $\overline{X}$ and R.

TABLE 18—*Operating characteristic, lot by lot control data.*

Lot	Sample Size, n	Average, $\overline{X}$	Range, R
No. 1	5	36.0	6.6
No. 2	5	31.4	0.5
No. 3	5	39.0	15.1
No. 4	5	35.6	8.8
No. 5	5	38.8	2.2
No. 6	5	41.6	3.5
No. 7	5	36.2	9.6
No. 8	5	38.0	9.0
No. 9	5	31.4	20.6
No. 10	5	29.2	21.7

Central Lines

For $\overline{X}$: $\overline{X}_0' = 35.00$

$n = 5$

For R: $d_2\sigma_0' = 2.326\,(4.20) = 9.8$

Control Limits

$n = 5$

For $\overline{X}$: $\overline{X}_0' \pm A\sigma_0' =$
$35.00 \pm (1.342)\,(4.20)$,
40.6, and 29.4

For R: $D_2\sigma_0'$ and $D_1\sigma_0' =$
$(4.918)\,(4.20)$ and $(0)\,(4.20)$,
20.7, and 0

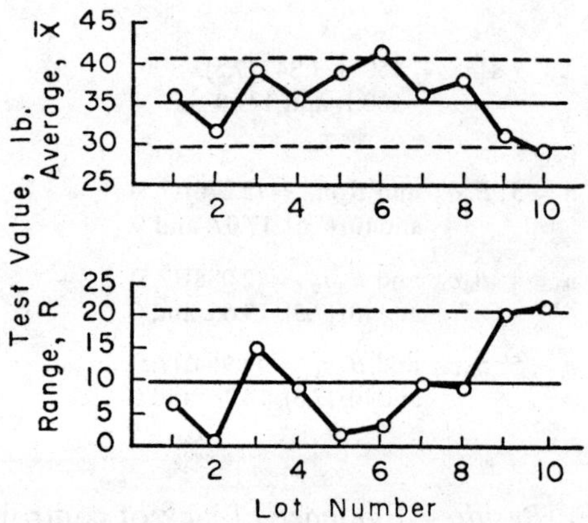

FIG. 20—*Control charts for* X *and* R. *Small samples of equal size,* n = 5; X$_0'$, σ_0' *given.*

Results—Lack of control at the standard level is indicated by results for lots Numbers 6 and 10. Corrective action is required both with respect to averages and with respect to variability within a lot.

Example 17: Control Charts for p, *Samples of Equal Size (Section 23) and* pn, *Samples of Equal Size (Section 24)*

Consider the same data as in Example 7, Table 9. The manufacturer wishes to control his process with respect to finish on galvanized washers at a level such that the fraction defective $p' = 0.0040$ (4 defective washers per thousand). Table 9 of Example 7 gives observed values of "number of defectives" for finish defects such as exposed steel, rough galvanizing in samples of 400 washers drawn from 15 successive lots. Figure 21 shows the control chart for p, and Fig. 22 gives the control chart for pn. In practice, only one of these control charts would be used since, except for change of scale, the two charts are identical.

(*a*) p

Central Line

$$p_0' = 0.0040$$

Control Limits

$n = 400$

$$p_0' \pm 3\sqrt{\frac{p_0'(1 - p_0')}{n}} =$$

$$0.0040 \pm 3\sqrt{\frac{0.0040\,(0.9960)}{400}} =$$

$$0.0040 \pm 0.0095,\ 0.0135,\ \text{and}\ 0$$

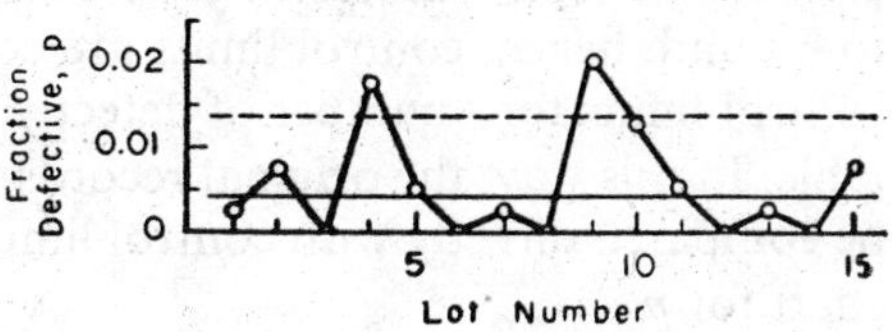

FIG. 21—*Control chart for* p. *Samples of equal size,* n = *400;* p_0' *given.*

(*b*) pn

Central Line

$$p_0'n = 0.0040\,(400) = 1.6$$

Control Limits

Exact formula:

$n = 400$

$$p_0'n \pm 3\sqrt{p_0'n(1 - p_0')} =$$

$$1.6 \pm 3\sqrt{1.6\,(0.996)} =$$

$$1.6 \pm 3\sqrt{1.5936} =$$

$$1.6 \pm 3(1.262),$$

$$5.4,\ \text{and}\ 0$$

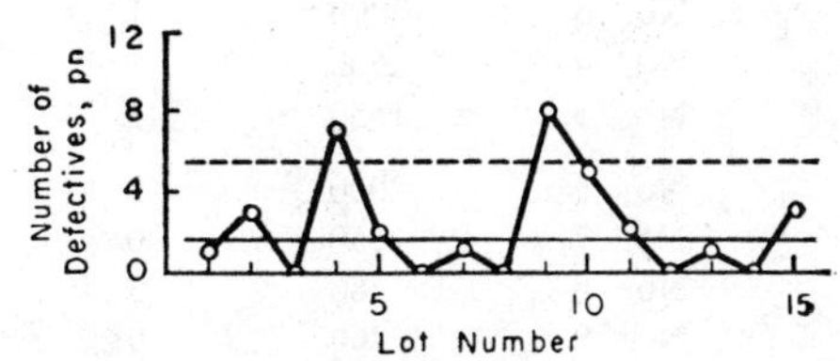

FIG. 22—*Control chart for* pn. *Samples of equal size,* n = *400;* p_0' *given.*

Simplified approximate formula:

$n = 400$

Since p_0' is small, replace Eq 29 by Eq 30

$$p_0'n \pm 3\sqrt{p_0'n} =$$

$$1.6 \pm 3\sqrt{1.6} =$$

$$1.6 \pm 3(1.265),\ 5.4,\ \text{and}\ 0$$

Results—Lack of control of quality is indicated with respect to the desired level; lots Numbers 4 and 9 are outside control limits.

Example 18: Control Chart for p *(Fraction Defective), Samples of Unequal Size (Section 23)*

The manufacturer wished to control the quality of a type of electrical apparatus with respect to two adjustment characteristics at a level such that the fraction defective $p' = 0.0020$ (2 defective units per thousand). Table 19 gives observed values of "number of defectives" for this item found in samples drawn from successive lots. Sample sizes vary considerably from lot to lot and, hence, control limits are computed for each sample. Equivalent control limits for "number of defectives," pn, are shown in column 5 of the table. In this way, the original records showing "number of defectives" may be compared directly with control limits for pn. Figure 23 shows the control chart for p.

TABLE 19—*Adjustment irregularities, electrical apparatus*

Lot	Sample Size, n	Number of Defectives	Fraction Defective, p	Upper Control limit for pn	Upper Control Limit for p
No. 1	600	2	0.0033	4.5	0.0075
No. 2	1300	2	0.0015	7.4	0.0057
No. 3	2000	1	0.0005	10.0	0.0050
No. 4	2500	1	0.0004	11.7	0.0047
No. 5	1550	5	0.0032	8.4	0.0054
No. 6	2000	2	0.0010	10.0	0.0050
No. 7	1550	0	0	8.4	0.0054
No. 8	780	3	0.0038	5.3	0.0068
No. 9	260	0	0	2.7	0.0103
No. 10	2000	15	0.0075	10.0	0.0050
No. 11	1550	7	0.0045	8.4	0.0054
No. 12	950	2	0.0021	6.0	0.0063
No. 13	950	5	0.0053	6.0	0.0063
No. 14	950	2	0.0021	6.0	0.0063
No. 15	35	0	0	0.9	0.0247
No. 16	330	3	0.0091	3.1	0.0094
No. 17	200	0	0	2.3	0.0115
No. 18	600	4	0.0067	4.5	0.0075
No. 19	1300	8	0.0062	7.4	0.0057
No. 20	780	4	0.0051	5.3	0.0068

Central Line for p

$p_0' = 0.0020$

Control Limits for p

$$p_0' \pm 3\sqrt{\frac{p_0'(1 - p_0')}{n}}$$

For $n = 600$:

$$0.0020 \pm 3\sqrt{\frac{0.002\,(0.998)}{600}} =$$

0.0020 ± 3(0.001824), 0.0075, and 0

(same procedure for other values of n)

Control Limits for pn

Using Eq 30 for pn,

$$p_0'n \pm 3\sqrt{p_0'n}$$

For $n = 600$:

$$1.2 \pm 3\sqrt{1.2} = 1.2 \pm 3\,(1.095),$$

4.5, and 0

(same procedure for other values of n)

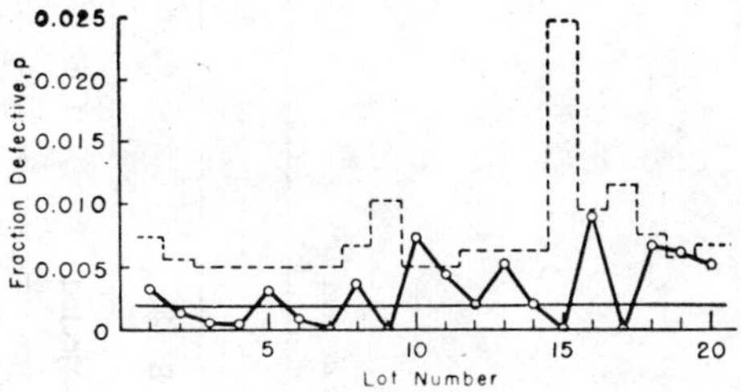

FIG. 23—*Control chart for* p. *Samples of unequal size,* n = *35 to 2500;* p_0' *given.*

Results—Lack of control and need for corrective action indicated by results for lots Numbers 10 and 19.

Example 19: Control Chart for p *(Fraction Rejected), Total and Components, Samples of Unequal Size (Section 23)*

A control device was given a 100 percent inspection in lots varying in size from about 1800 to 5000 units, each unit being tested and inspected with respect to 23 *essentially independent* characteristics. These 23 characteristics were grouped into three groups designated Groups A, B, and C, corresponding to three successive inspections. A unit found defective at any time with respect to any one characteristic was immediately rejected; hence units found defective in, say, the Group A inspection were not subjected to the two subsequent group inspections. In fact, the number of units inspected for each characteristic in a group itself will differ from characteristic to to characteristic if defects with respect to the characteristics in a group occur, the last characteristic in the group having the smallest sample size. Since 100 percent inspection is used, no additional units are available for inspection to maintain a constant sample size for all characteristics in a group or for

TABLE 20—*Inspection data for 100 percent inspection—control device.*

	Observed Number of Rejects and Fraction Rejected											
	All Groups Combined			Group A			Group B			Group C		
Lot	Lot Size, n	Total Rejected Number	Total Rejected Fraction	Lot Size, n	Rejected Number	Rejected Fraction	Lot Size, n	Rejected Number	Rejected Fraction	Lot Size, n	Rejected Number	Rejected Fraction
No. 1	4814	914	0.190	4814	311	0.065	4503	253	0.056	4250	350	0.082
No. 2	2159	359	0.166	2159	128	0.059	2031	105	0.052	1926	126	0.065
No. 3	3089	565	0.183	3089	195	0.063	2894	149	0.051	2745	221	0.081
No. 4	3156	626	0.198	3156	233	0.074	2923	142	0.049	2781	251	0.090
No. 5	2139	434	0.203	2139	146	0.068	1993	101	0.051	1892	187	0.099
No. 6	2588	503	0.194	2588	177	0.068	2411	151	0.063	2260	175	0.077
No. 7	2510	487	0.194	2510	143	0.057	2367	116	0.049	2251	228	0.101
No. 8	4103	803	0.196	4103	318	0.078	3785	242	0.064	3543	243	0.069
No. 9	2992	547	0.183	2992	208	0.070	2784	130	0.047	2654	209	0.079
No. 10	3545	643	0.181	3545	172	0.049	3373	180	0.053	3193	291	0.091
No. 11	1841	353	0.192	1841	97	0.053	1744	119	0.068	1625	137	0.084
No. 12	2748	418	0.152	2748	141	0.051	2607	114	0.044	2493	163	0.065

Central Lines and Control Limits, Based on Standard p'_0 Values

	All Groups Combined	Group A	Group B	Group C
	Central Lines			
p'_0 =	0.180	0.070	0.050	0.080
Lot	Control Limits			
No. 1	0.197 and 0.163	0.081 and 0.059	0.060 and 0.040	0.093 and 0.067
No. 2	0.205 and 0.155	0.086 and 0.054	0.064 and 0.036	0.099 and 0.061
No. 3	0.201 and 0.159	0.084 and 0.056	0.062 and 0.038	0.096 and 0.064
No. 4	0.200 and 0.160	0.084 and 0.056	0.062 and 0.038	0.095 and 0.065
No. 5	0.205 and 0.155	0.086 and 0.054	0.065 and 0.035	0.099 and 0.061
No. 6	0.203 and 0.157	0.085 and 0.055	0.063 and 0.037	0.097 and 0.063
No. 7	0.203 and 0.157	0.085 and 0.055	0.064 and 0.036	0.097 and 0.063
No. 8	0.198 and 0.162	0.082 and 0.058	0.061 and 0.039	0.094 and 0.066
No. 9	0.201 and 0.159	0.084 and 0.056	0.062 and 0.038	0.096 and 0.064
No. 10	0.200 and 0.160	0.083 and 0.057	0.061 and 0.039	0.094 and 0.066
No. 11	0.207 and 0.153	0.088 and 0.052	0.066 and 0.034	0.100 and 0.060
No. 12	0.202 and 0.158	0.085 and 0.055	0.063 and 0.037	0.096 and 0.064

all the component groups. The fraction defective with respect to each characteristic is sufficiently small so that the error within a group, although rather large between the first and last characteristic inspected by one inspection group, can be neglected for practical purposes. Under these circumstances, the number inspected for any group was equal to the lot size diminished by the number of units rejected in the preceding inspections.

Part 1 of Table 20 gives the data for 12 successive lots of product, and shows for each lot inspected the total fraction rejected as well as the number and fraction rejected at each inspection station. Part 2 of Table 20 gives values of p_0', fraction rejected, at which levels the manufacturer desires to control this device, with respect to all twenty-three characteristics combined and with respect to the characteristics tested and inspected at each of the three inspection stations. Note that the p_0' for all characteristics (in terms of defectives) is less than the sum of the p_0' values for the three component groups, since defects from more than one characteristic or group of characteristics may occur on a single unit. Control limits, lower and upper, in terms of fraction rejected are listed for each lot size using the initial lot size as the sample size for all characteristics combined and the lot size available at the beginning of inspection and test for each group as the sample size for that group.

Figure 24 shows four control charts, one covering all rejections combined for the control device and three other charts covering the rejections for each

Central Lines
See Table 20

Control Limits
See Table 20
For Lot Number 1

Total: $n = 4814$

$$p_0' \pm 3\sqrt{\frac{p_0'(1 - p_0')}{n}} =$$

$$0.180 \pm 3\sqrt{\frac{0.180\,(0.820)}{4814}} =$$

$0.180 \pm 3(0.0055)$, 0.197, and 0.163

Group C: $n = 4250$

$$p_0' \pm 3\sqrt{\frac{p_0'(1 - p_0')}{n}} =$$

$$0.080 \pm 3\sqrt{\frac{0.080\,(0.920)}{4250}} =$$

$0.080 \pm 3(0.0042)$, 0.093, and 0.067

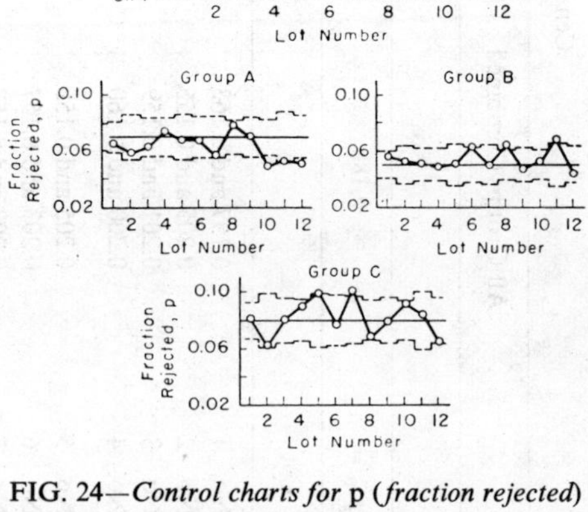

FIG. 24—*Control charts for* p *(fraction rejected) for total and components. Samples of unequal size,* n = *1625 to 4814;* p_0' *given.*

of the three inspection stations for Group A, Group B, and Group C characteristics, respectively. Detailed computations for the over-all results for one lot and one of its component groups are given.

Results—Lack of control is indicated for all characteristics combined; lot Number 12 is outside control limits in a favorable direction and the corresponding results for each of the three components are less than their standard values, Group A being below the lower control limit. For Group A results, lack of control is indicated since lot Numbers 10 and 12 are below their lower control limits. Lack of control is indicated for the component characteristics in Group B, since lot Numbers 8 and 11 are above their upper control limits. For Group C, lot Number 7 is above its upper limit indicating lack of control. Corrective measures are indicated for Groups B and C and steps should be taken to determine whether the Group A component might not be controlled at a smaller value of p_0', such as 0.06.

Example 20: Control Chart for u, *Samples of Unequal Size* (*Section 25*)

It is desired to control the number of defects per billet to a standard of 1.000 defects per unit in order that the wire made from such billets of copper will not contain an excessive number of defects. The lot sizes varied greatly from day to day so that a sampling schedule was set up giving three different sample sizes to cover the range of lot sizes received. A control program was instituted using a control chart for defects per unit with reference to the desired standard. Table 21 gives data in terms of defects and defects per unit for 15 consecutive lots under this program. Figure 25 shows the control chart for u.

TABLE 21—*Lot by lot inspection results for copper billets in terms of defects and defects per unit.*

Lot	Sample Size, n	Number of Defects, c	Defects per Unit, u	Lot	Sample Size, n	Number of Defects, c	Defects per Unit, u
No. 1	100	75	0.750	No. 10	100	130	1.300
No. 2	100	138	1.380	No. 11	100	58	0.580
No. 3	200	212	1.060	No. 12	400	480	1.200
No. 4	400	444	1.110	No. 13	400	316	0.790
No. 5	400	508	1.270	No. 14	200	162	0.810
No. 6	400	312	0.780	No. 15	200	178	0.890
No. 7	200	168	0.840				
No. 8	200	266	1.330	Total	3500	3566	
No. 9	100	119	1.190	Overall[a]			1.019

[a] $\bar{u} = 3566/3500 = 1.019$.

Central Line
$u_0' = 1.000$

Control Limits
$n = 100$

$$u_0' \pm 3\sqrt{\frac{u_0'}{n}} =$$

$$1.000 \pm 3\sqrt{\frac{1.000}{100}} =$$

$1.000 \pm 3(0.100)$, 1.300, and 0.700

$n = 200$

$$u_0' \pm 3\sqrt{\frac{u_0'}{n}} =$$

$$1.000 \pm 3\sqrt{\frac{1.000}{200}} =$$

$1.000 \pm 3(0.0707)$, 1.212, and 0.788

$n = 400$

$$u_0' \pm 3\sqrt{\frac{u_0'}{n}} =$$

$$1.000 \pm 3\sqrt{\frac{1.000}{400}} =$$

$1.000 \pm 3(0.0500)$, 1.150, and 0.850

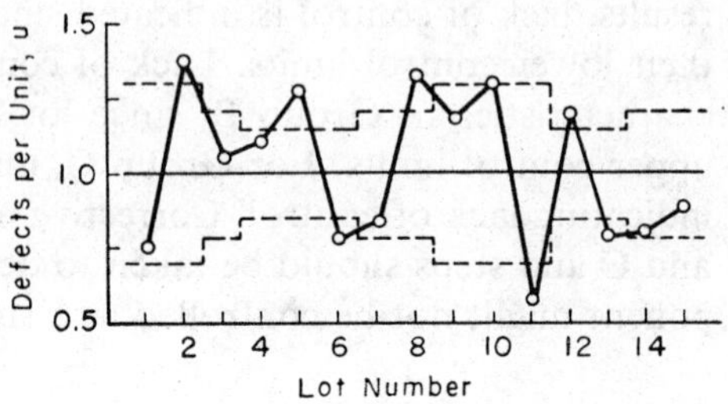

FIG. 25—*Control chart for* u. *Samples of unequal size,* n = *100, 200, 400;* u_0' *given.*

Results—Lack of control of quality is indicated with respect to the desired level since lot Numbers 2, 5, 8, and 12 are above the upper control limit and lot Numbers 6, 11, and 13 are below the lower control limit. The over-all level, 1.019 defects per unit, is slightly above the desired value of 1.000 defects defects per unit. Corrective action is necessary to reduce the spread between successive lots and reduce the average number of defects per unit.

Example 21: Control Charts for c, *Samples of Equal Size* (*Section 26*)

A Type D motor is being produced by a manufacturer and he desires to control the number of defects per motor at a level of $u' = 3.000$ defects per unit with respect to all visual defects. He is producing on a continuous basis and decides to take a sample of 25 motors every day, where a day's product is treated as a lot. Due to the nature of the process, he plans on controlling the product for these defects at a level such that $c_0' = 75.0$ defects as $u_0'n = c_0'$. Table 22 gives data in terms of number of defects, c, and also defects per unit,

u, for 10 consecutive days. Figure 26 shows the control chart for c. As in Example 20, a control chart may be made for u, where the central line is $u_0' = 3.000$ and the control limits are

$$u_0' \pm 3\sqrt{\frac{u_0'}{n}} = 3.000 \pm 3\sqrt{\frac{3.000}{25}} = 3.000 \pm 3(0.3464), 4.04, \text{ and } 1.96$$

TABLE 22—*Daily inspection results for Type D motors in terms of defects per sample and defects per unit.*

Lot	Sample Size, n	Number of Defects, c	Defects per Unit, u
No. 1	25	81	3.24
No. 2	25	64	2.56
No. 3	25	53	2.12
No. 4	25	95	3.80
No. 5	25	50	2.00
No. 6	25	73	2.92
No. 7	25	91	3.64
No. 8	25	86	3.44
No. 9	25	99	3.96
No. 10	25	60	2.40
Total	250	752	30.08
Average		75.2	3.008

Central Line

$$c_0' = u_0'n = 3.000 \times 25 = 75.0$$

Control Limits
$n = 25$

$$c_0' \pm 3\sqrt{c_0'} =$$

$$75.0 \pm 3\sqrt{75.0} =$$
$$75.0 \pm 3(8.66), 100.98, \text{ and } 49.02$$

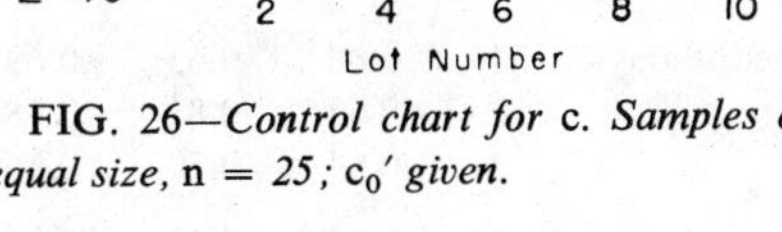

FIG. 26—*Control chart for* c. *Samples of equal size,* n = *25;* c₀′ *given.*

Results—No significant deviations from the desired level.

33. Illustrative Examples—Control Chart for Individuals

Examples 22 to 35, inclusive, illustrate the use of the control chart for individuals, in which individual observations are plotted one by one. The examples cover the two general conditions: (*a*) control, no standard given; and (*b*) control with respect to a given standard (see Sections 28 to 30).

Example 22: Control Chart for Individuals, X—*Using Rational Subgroups, Samples of Equal Size, No Standard Given—Based on* $\overline{\mathrm{X}}$ *and* $\overline{\mathrm{R}}$ *(Section 29)*

In the manufacture of manganese steel tank shoes, five 4-ton heats of metal were cast in each 8-h shift, the silicon content being controlled by ladle additions computed from preliminary analyses. A high silicon content was known to aid in the production of sound castings, but the specification set a maximum of 1.00 percent silicon for a heat, and all shoes from a heat exceeding this specification were rejected. It was important, therefore, to detect any trouble with silicon control before even one heat exceeded the specification.

Since the heats of metal were well stirred, within-heat variation of silicon content was not a useful basis for control limits. However, each 8-h shift used the same materials, equipment etc., and the quality depended largely on the care and efficiency with which they operated so that the five heats produced in an 8-h shift provided a rational subgroup..

Data analyzed in the course of an investigation and before standard values were established are shown in Table 23 and control charts for $\overline{X}$, R, and X are shown in Fig. 27.

TABLE 23—*Silicon content of heats of manganese steel, percent*

Day	Shift	Heat 1	2	3	4	5	Sample Size, n	Average, $\overline{X}$	Range, R
Monday	1	0.70	0.72	0.61	0.75	0.73	5	0.702	0.14
	2	0.83	0.68	0.83	0.71	0.73	5	0.756	0.15
	3	0.86	0.78	0.71	0.70	0.90	5	0.790	0.20
Tuesday	1	0.80	0.78	0.68	0.70	0.74	5	0.740	0.12
	2	0.64	0.66	0.79	0.81	0.68	5	0.716	0.17
	3	0.68	0.64	0.71	0.69	0.81	5	0.706	0.17
Wednesday	1	0.80	0.63	0.69	0.62	0.75	5	0.698	0.18
	2	0.65	0.81	0.68	0.84	0.66	5	0.728	0.19
	3	0.64	0.70	0.66	0.65	0.93	5	0.716	0.29
Thursday	1	0.77	0.83	0.88	0.70	0.64	5	0.764	0.24
	2	0.72	0.67	0.77	0.74	0.72	5	0.724	0.10
	3	0.73	0.66	0.72	0.73	0.71	5	0.710	0.07
Friday	1	0.79	0.70	0.63	0.70	0.88	5	0.740	0.25
	2	0.85	0.80	0.78	0.85	0.62	5	0.780	0.23
	3	0.67	0.78	0.81	0.84	0.96	5	0.812	0.29
Total	15							11.082	2.79
Average								0.7388	0.186

Central Lines

For $\bar{X}$: $\bar{\bar{X}} = 0.7388$
For R: $\bar{R} = 0.186$
For X: $\bar{X} = 0.7388$

Control Limits
$n = 5$

For $\bar{X}$: $\bar{\bar{X}} \pm A_2\bar{R} =$
$0.7388 \pm (0.577)(0.186)$,
0.846, and 0.631

For R: $D_4\bar{R}$ and $D_3\bar{R} =$
$(2.115)(0.186)$ and $(0)(0.186)$,
0.393, and 0

For X: $\bar{X} \pm E_2\bar{R} =$
$0.7388 \pm (1.290)(0.186)$,
0.979, and 0.499.

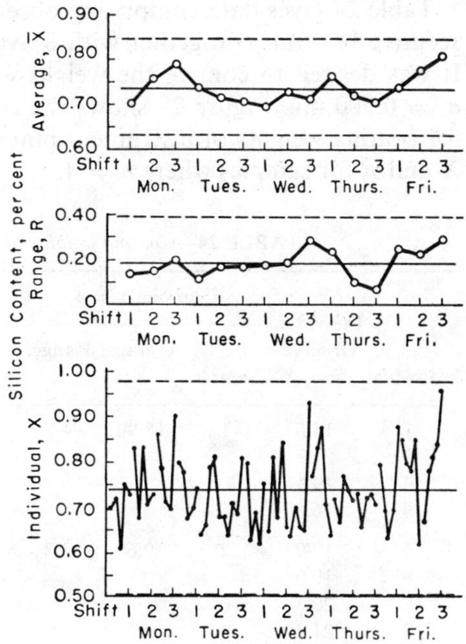

FIG. 27—*Control charts for* $\bar{X}$, R, *and* X. *Samples of equal size,* n = 5; *no standard given.*

Results—None of the charts give evidence of lack of control.

Example 23: Control Chart for Individuals, X—*Using Rational Subgroups, Standard Given, Based on* $\bar{X}_0'$ *and* σ_0' (*Section 29*)

In the hand spraying of small instrument pins held in bar frames of 25 each, when coating thickness and weight had to be delicately controlled, spray-gun adjustments were critical and had to be watched continuously from bar to bar. Weights were measured differently by careful weighing before and after removal of the coating. Destroying more than one pin per bar was economically not feasible, yet failure to catch a bar departing from standards might result in the unsatisfactory performance of some 24 assembled instruments. The standard lot size for these instrument pins was 100 so that initially control charts for average and range were set up with $n = 4$. It was found that the variation in thickness of coating on the 25 pins on a single bar was quite small as compared with the between-bar variation. Accordingly, as an adjunct to the control charts for average and range, a control chart for individuals, X, at the sprayer position was adopted for the operator's guidance.

Table 24 gives data comprising observations on 32 pins taken from consecutive bar frames together with 8 average and range values were $n = 4$. It was desired to control the weight with an average $\bar{X}' = 20.00$ mg and $\sigma' = 0.900$ mg. Figure 28 shows the control chart for individual values X for coating weights of instrument pins together with the control charts for $\bar{X}$ and R for samples where $n = 4$.

TABLE 24—*Coating weights of instrument pins, milligrams.*

Individual	Individual Observation, X	Sample, $n = 4$: Sample	Average, $\bar{X}$	Range, R	Individual	Individual Observation, X	Sample, $n = 4$: Sample	Average, $\bar{X}$	Range, R
1	18.5	1	18.90	4.7	17	19.1	5	20.52	2.5
2	21.2				18	20.6			
3	19.4				19	20.8			
4	16.5				20	21.6			
5	17.9	2	19.60	3.3	21	22.8	6	22.80	1.0
6	19.0				22	22.2			
7	20.3				23	23.2			
8	21.2				24	23.0			
9	19.6	3	20.08	0.9	25	19.0	7	19.75	1.5
10	19.8				26	20.5			
11	20.4				27	20.3			
12	20.5				28	19.2			
13	22.2	4	21.20	1.9	29	20.7	8	20.32	1.9
14	21.5				30	21.0			
15	20.8				31	20.5			
16	20.3				32	19.1			
					Total	652.7		163.17	17.7
					Average	20.40		20.40	2.21

Results—All three charts show lack of control. At the outset both the chart for ranges and the chart for individuals gave indications of lack of control. Subsequently for Sample 6 the control chart for individuals showed the first unit in the sample of 4 to be outside its upper control limit, thus indicating lack of control before the entire sample was obtained.

Example 24: Control Charts for Individuals, X, *and Moving Range,* R, *of Two Observations, No Standard Given—Based on* $\bar{X}$ *and* $\bar{R}$, *the Mean Moving Range* (*Section 30*(a))

A distilling plant was distilling and blending batch lots of denatured alcohol in a large tank. It was desired to control the percentage of methanol

(Example 23—Continued)

Central Line

For X: $\bar{X}_0' = 20.00$

Control Limits

For X: $\bar{X}_0' \pm 3\sigma_0' =$
$20.00 \pm 3(0.900)$,
22.7, and 17.3

Central Lines

For $\bar{X}$: $\bar{X}_0' = 20.00$
For R: $\bar{R}' = d_2\sigma_0' =$
(2.059) (0.900)
1.85

Control Limits
$n = 4$

For $\bar{X}$: $\bar{X}_0' \pm A\sigma_0' =$
$20.00 \pm (1.500)(0.900)$,
21.35, and 18.65

For R: $D_2\sigma_0'$ and $D_1\sigma_0' =$
(4.698) (0.900) and
(0) (0.900),
4.23, and 0

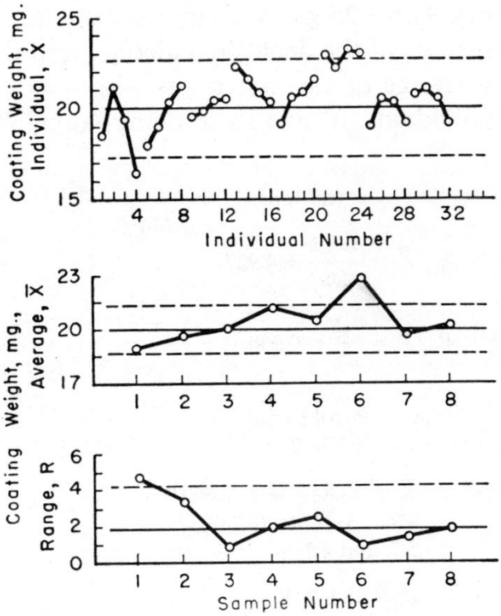

FIG. 28—*Control charts for* X, $\bar{X}$, *and* R. *Small samples of equal size,* n = *4;* $\bar{X}_0'$, σ_0' *given.*

TABLE 25—*Methanol content of successive lots of denatured alcohol and moving range for* n = 2.

Lot	Percentage of Methanol, X	Moving Range, R	Lot	Percentage of Methanol, X	Moving Range, R
No. 1	4.6	...	No. 14	5.5	0.1
No. 2	4.7	0.1	No. 15	5.2	0.3
No. 3	4.3	0.4	No. 16	4.6	0.6
No. 4	4.7	0.4	No. 17	5.5	0.9
No. 5	4.7	0	No. 18	5.6	0.1
No. 6	4.6	0.1	No. 19	5.2	0.4
No. 7	4.8	0.2	No. 20	4.9	0.3
No. 8	4.8	0	No. 21	4.9	0
No. 9	5.2	0.4	No. 22	5.3	0.4
No. 10	5.0	0.2	No. 23	5.0	0.3
No. 11	5.2	0.2	No. 24	4.3	0.7
No. 12	5.0	0.2	No. 25	4.5	0.2
No. 13	5.6	0.6	No. 26	4.4	0.1
			Total	128.1	0.2

for this process. The variability of sampling within a single lot was found to be negligible so it was decided feasible to take only one observation per lot and to set control limits on the basis of the moving range of successive lots. Table 25 gives a summary of the methanol content, X, of 26 consecutive lots of the denatured alcohol and the 25 values of the moving range, R, the range of successive lots with $n = 2$. Figure 29 gives control charts for individuals, X, and the moving range, R.

Central Lines

For X: $\bar{X} = \dfrac{128.1}{26} = 4.927$

For R: $\bar{R} = \dfrac{7.2}{25} = 0.288$

Control Limits
$n = 2$

For X: $\bar{X} \pm E_2\bar{R} = \bar{X} \pm 2.660\bar{R} =$
$4.927 \pm (2.660)(0.288)$,
5.7, and 4.2

For R: $D_4\bar{R}$ and $D_3\bar{R} =$
(3.267)(0.288) and
(0)(0.288), 0.94, and 0

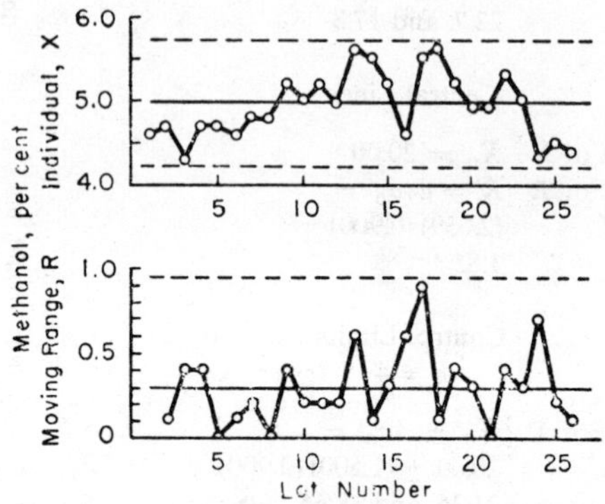

FIG. 29—*Control charts for* X *and* R. *No standard given; based on moving range, where* n = 2.

Results—The trend pattern of the individuals and their tendency to crowd the control limits suggests that better control may be attainable.

Example 25: Control Charts for Individuals, X, *and Moving Range,* R, *of Two Observations, Standard Given—Based on* $\bar{X}_0'$ *and* σ_0' (*Section 30*(b))

Data from same source as for Example 24 in which a distilling plant was distilling and blending batch lots of denatured alcohol in a large tank. It was desired to control the percentage of water for this process. The variability of sampling within a single lot was found to be negligible so it was decided to take only one observation per lot and to set control limits for individual values, X, and for the moving range, R, of successive lots with $n = 2$ where $\bar{X}_0' = 7.800$ percent and $\sigma_0' = 0.200$ percent. Table 26 gives a summary of the water content of 26 consecutive lots of the denatured alcohol and the 25 values of the moving range, R. Figure 30 gives control charts for individuals, X, and for the moving range, R.

TABLE 26—*Water content of successive lots of denatured alcohol and moving range for* n = 2.

Lot	Percentage of Water, X	Moving Range, R	Lot	Percentage of Water, X	Moving Range, R
No. 1	8.9	...	No. 14	8.2	0.3
No. 2	7.7	1.2	No. 15	8.2	0
No. 3	8.2	0.5	No. 16	7.5	0.7
No. 4	7.9	0.3	No. 17	7.5	0
No. 5	8.0	0.1	No. 18	7.8	0.3
No. 6	8.0	0	No. 19	8.5	0.7
No. 7	7.7	0.3	No. 20	7.5	1.0
No. 8	7.8	0.1	No. 21	8.0	0.5
No. 9	7.9	0.1	No. 22	8.5	0.5
No. 10	8.2	0.3	No. 23	8.4	0.1
No. 11	7.5	0.7	No. 24	7.9	0.5
No. 12	7.5	0	No. 25	8.4	0.5
No. 13	7.9	0.4	No. 26	7.5	0.9
			Total	207.1	10.0
			Number of values	26	25
			Average	7.965	0.400

Central Lines

For X: $\overline{X}_0' = 7.800$

$n = 2$

For R: $d_2\sigma_0' = (1.128)(0.200)$, 0.23

Control Limits

For X: $\overline{X}_0' \pm 3\sigma' =$ 7.800 $\pm$ 3(0.200), 8.4, and 7.2

$n = 2$

For R: $D_2\sigma_0'$ and $D_1\sigma_0'$ = (3.686)(0.200) and (0)(0.200), 0.74, and 0

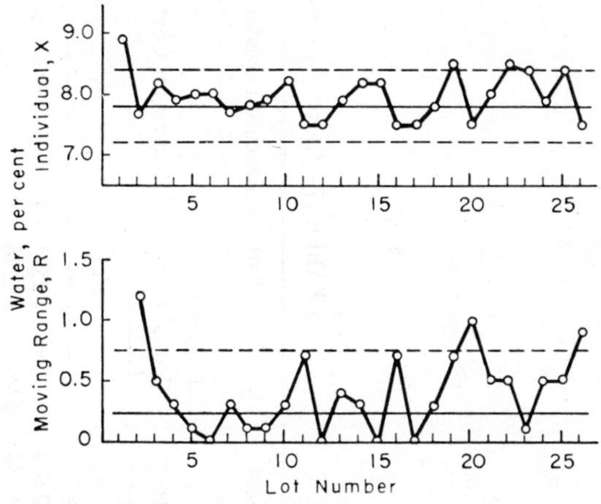

FIG. 30—*Control charts for* X *and moving range,* R, *where* n = 2. *Standard given; based on* $\overline{X}_0'$ *and* σ_0'.

Results—Lack of control at desired levels is indicated with respect to both the individual readings and the moving range. These results indicate corrective measures should be taken to reduce the level in percent and to reduce the variation between lots.

TABLE 27—*Factors for computing control chart lines.*

Observations in Sample, n	Chart for Averages			Chart for Standard Deviations						Chart for Ranges						
	Factors for Control Limits			Factors for Central Line		Factors for Control Limits				Factors for Central Line		Factors for Control Limits				
	A	A_2	A_3	c_4	$1/c_4$	B_3	B_4	B_5	B_6	d_2	$1/d_2$	d_3	D_1	D_2	D_3	D_4
2	2.121	1.880	2.659	0.7979	1.2533	0	3.267	0	2.606	1.128	0.8865	0.853	0	3.686	0	3.267
3	1.732	1.023	1.954	0.8862	1.1284	0	2.568	0	2.276	1.693	0.5907	0.888	0	4.358	0	2.574
4	1.500	0.729	1.628	0.9213	1.0854	0	2.266	0	2.088	2.059	0.4857	0.880	0	4.698	0	2.282
5	1.342	0.577	1.427	0.9400	1.0638	0	2.089	0	1.964	2.326	0.4299	0.864	0	4.918	0	2.114
6	1.225	0.483	1.287	0.9515	1.0510	0.030	1.970	0.029	1.874	2.534	0.3946	0.848	0	5.078	0	2.004
7	1.134	0.419	1.182	0.9594	1.0423	0.118	1.882	0.113	1.806	2.704	0.3698	0.833	0.204	5.204	0.076	1.924
8	1.061	0.373	1.099	0.9650	1.0363	0.185	1.815	0.179	1.751	2.847	0.3512	0.820	0.388	5.306	0.136	1.864
9	1.000	0.337	1.032	0.9693	1.0317	0.239	1.761	0.232	1.707	2.970	0.3367	0.808	0.547	5.393	0.184	1.816
10	0.949	0.308	0.975	0.9727	1.0281	0.284	1.716	0.276	1.669	3.078	0.3249	0.797	0.687	5.469	0.223	1.777
11	0.905	0.285	0.927	0.9754	1.0252	0.321	1.679	0.313	1.637	3.173	0.3152	0.787	0.811	5.535	0.256	1.744
12	0.866	0.266	0.886	0.9776	1.0229	0.354	1.646	0.346	1.610	3.258	0.3069	0.778	0.922	5.594	0.283	1.717
13	0.832	0.249	0.850	0.9794	1.0210	0.382	1.618	0.374	1.585	3.336	0.2998	0.770	1.025	5.647	0.307	1.693
14	0.802	0.235	0.817	0.9810	1.0194	0.406	1.594	0.399	1.563	3.407	0.2935	0.763	1.118	5.696	0.328	1.672
15	0.775	0.223	0.789	0.9823	1.0180	0.428	1.572	0.421	1.544	3.472	0.2880	0.756	1.203	5.741	0.347	1.653

16	0.750	0.212	0.763	0.9835	1.0168	0.448	1.552	0.440	1.526	3.532	0.2831	0.750	1.282	5.782	0.363	1.637
17	0.728	0.203	0.739	0.9845	1.0157	0.466	1.534	0.458	1.511	3.588	0.2787	0.744	1.356	5.820	0.378	1.622
18	0.707	0.194	0.718	0.9854	1.0148	0.482	1.518	0.475	1.496	3.640	0.2747	0.739	1.424	5.856	0.391	1.608
19	0.688	0.187	0.698	0.9862	1.0140	0.497	1.503	0.490	1.483	3.689	0.2711	0.734	1.487	5.891	0.403	1.597
20	0.671	0.180	0.680	0.9869	1.0133	0.510	1.490	0.504	1.470	3.735	0.2677	0.729	1.549	5.921	0.415	1.585
21	0.655	0.173	0.663	0.9876	1.0126	0.523	1.477	0.516	1.459	3.778	0.2647	0.724	1.605	5.951	0.425	1.575
22	0.640	0.167	0.647	0.9882	1.0119	0.534	1.466	0.528	1.448	3.819	0.2618	0.720	1.659	5.979	0.434	1.566
23	0.626	0.162	0.633	0.9887	1.0114	0.545	1.455	0.539	1.438	3.858	0.2592	0.716	1.710	6.006	0.443	1.557
24	0.612	0.157	0.619	0.9892	1.0109	0.555	1.445	0.549	1.429	3.895	0.2567	0.712	1.759	6.031	0.451	1.548
25	0.600	0.135	0.606	0.9896	1.0105	0.565	1.435	0.559	1.420	3.931	0.2544	0.708	1.806	6.056	0.459	1.541
Over 25	$3/\sqrt{n}$	...	...	...	...	...[a]	...[b]	...[a]	...[b]	...	...	...	...	...	...	...

NOTES—

To conform with the change in the definition of the sample standard deviation, the c_2 column and the $1/c_2$ column were replaced in 1975 by the c_4 column and $1/c_4$ column, respectively. Likewise, the A_1 column was deleted and the A_3 column was added. Similarly, the E_1 column of Table 28 was deleted and the E_3 column was added.

Values of A_3, c_4, B_5, and B_6 have been reproduced by permission from the American Society for Quality Control (ASQC) Standard Al-1971 (ANSI Standard Z1.5-1971) Definitions, Symbols, Formulas, and Tables for Control Charts. Values of $1/c_4$ and E_3 were computed by the 1975 Task Group.

In 1975, values for $n = 14$ to 19 were corrected to equal those given in *Biometrika Tables for Statisticians*, Vol. 1, Table 20, by permission from the Biometrika Trust. The d_3 value for $n = 25$ was recomputed from the variance of $W = R/\sigma'$ given by Harter, H. L. in *Order Statistics and Their Use in Testing and Estimation*, Vol. 1, 1969, p. 376. Values of D_1, D_2, D_3, and D_4 were recomputed using the d_3 values computed to four significant figures from the variances given by Harter. The new figures differ from the old only in the third decimal place.

Those who prefer to define the sample standard deviation as $\sqrt{\Sigma(x - \bar{x})^2/n}$ are referred to *ASTM STP 15C* for the values of the factors c_2, A_1, B_1, B_2, and E_1 that are used when this definition is adopted. All but E_1 may also be found in ANSI standard Z1.5-1971.

[a] $1 - 3/\sqrt{2(n-1)}$.

[b] $1 + 3/\sqrt{2(n-1)}$.

Supplement A

Mathematical Relations and Tables of Factors for Computing Control Chart Lines

Scope—Supplement A presents mathematical relations used in arriving at the factors and formulas of Part 3. In addition, a more comprehensive tabulation of values of these factors is given in Tables 27 and 28, including reciprocal values of c_4 and d_2 and values of d_3. This last factor is involved in the relations covering control charts for ranges.

TABLE 28—*Factors for computing control lines—chart for individuals.*[a]

Observations in Sample, n	Chart for Individuals	
	Factors for Control Limits	
	E_2	E_3
2	2.660	3.760
3	1.772	3.385
4	1.457	3.256
5	1.290	3.191
6	1.184	3.153
7	1.109	3.129
8	1.054	3.109
9	1.010	3.095
10	0.975	3.084
11	0.946	3.076
12	0.921	3.069
13	0.899	3.063
14	0.881	3.058
15	0.864	3.054
16	0.849	3.050
17	0.836	3.047
18	0.824	3.044
19	0.813	3.042
20	0.803	3.040
21	0.794	3.038
22	0.785	3.036
23	0.778	3.034
24	0.770	3.033
25	0.763	3.032
Over 25	$3/d_2$	3

[a]See Notes in Table 27.

Factors c_4, d_2, *and* d_3 (*values for* n = 2 *to* 5, *inclusive, in Table 27*)—The relations given for factors c_4, d_2, and d_3 are based on sampling from a universe having a Normal distribution (see Ref *1*, p. 184).

$$c_4 = \sqrt{\frac{2}{n-1}} \frac{\left(\frac{n-2}{2}\right)!}{\left(\frac{n-3}{2}\right)!} \tag{42}$$

where the symbol ! as used indicates a factorial, for example, $4! = 4 \cdot 3 \cdot 2 \cdot 1 = 24$. For the relation $(k/2)!$ – if k is even, $(k/2)! = (k/2) \cdot ((k-2)/2) \cdot ((k-4)/2) \cdots 1$, each number on the right-hand side being an integer; if k is odd, $(k/2)! = (k/2) \cdot ((k-2)/2) \cdot ((k-4)/2) \cdots 1/2 \cdot \sqrt{\pi}$, where $(-1/2)! = \sqrt{\pi}$ and $0! = 1$.

$$d_2 = \int_{-\infty}^{\infty} [1 - (1-\alpha_1)^n - \alpha_1{}^n]\, dx_1 \tag{43}$$

where $\alpha_1 = \frac{1}{\sqrt{2\pi}} \int_{-\infty}^{x_1} e^{-(x^2/2)}\, dx$, n = sample size.

$$d_3 = \left[2 \int_{-\infty}^{\infty} \int_{-0}^{x_1} [1 - \alpha_1{}^n - (1-\alpha_n)^n + (\alpha_1 - \alpha_n)^n]\, dx_1\, dx_n - d_2{}^2\right]^{1/2} \tag{44}$$

where $\alpha_1 = \frac{1}{\sqrt{2\pi}} \int_{-\infty}^{x_1} e^{-(x^2/2)}\, dx$, $\alpha_n = \frac{1}{\sqrt{2\pi}} \int_{-\infty}^{x_n} e^{-(x^2/2)}\, dx$.

n = sample size and d_2 = average range for a Normal law distribution with standard deviation equal to unity. (In his original paper, Tippett [*9*] used w for range and $\bar{w}$ for d_2.)

The relations just mentioned for c_4, d_2, and d_3 are exact when the original universe is Normal but this does not limit their use in practice. They may for most practical purposes be considered satisfactory for use in control chart work even though the universe is not Normal. Since the relations are involved and thus difficult to compute, values of c_4, d_2, and d_3 for $n = 2$ to 25, inclusive, are given in Table 27. All values listed in the table were computed to enough significant figures so that when rounded off in accordance with standard practices the last figure shown in the table was not in doubt.

Standard Deviations of $\bar{X}$, s, R, p, pn, u, *and* c—The standard deviations of $\bar{X}$, s, R, p, etc., used in setting 3-sigma control limits designated $\sigma_{\bar{X}}'$, σ_s', σ_R', σ_p', etc., in Part 3 are the standard deviations of the sampling distributions of $\bar{X}$, s, R, p, etc., for subgroups (samples) of size n. They are not the standard deviations which might be computed from the subgroup values of $\overline{X}$, s, R, p, etc., plotted on the control charts but are computed by formula from the quantities listed in Table 29.

TABLE 29—*Basis of standard deviations for control limits.*

Control Chart	Standard Deviation Used in Computing 3-Sigma Limits is Computed From: Control—No Standard Given	Control—Standard Given
$\overline{X}$	$\bar{s}$ or $\overline{R}$	σ'_0
s	$\bar{s}$ or $\overline{R}$	σ'_0
R	$\bar{s}$ or $\overline{R}$	σ'_0
p	$\bar{p}$	p'_0
pn	$\bar{p}n$	p'_0n
u	$\bar{u}$	u'_0
c	$\bar{c}$	c'_0

NOTE—$\bar{s}$, $\overline{R}$, etc., are computed averages of subgroup values; σ'_0, p'_0, etc., are standard values.

The standard deviations $\sigma_{\overline{X}}'$, σ_s' and σ_R' computed in this way are unaffected by any assignable causes of variation between subgroups. Consequently, the control charts derived from them will detect assignable causes of this type.

The relations in Eqs 45 to 57, inclusive, which follow, are all of the form

standard deviation of the sampling distribution = a function of both the sample size, n, and a universe value σ', p', u', or c'

In practice, a sample estimate or standard value is substituted for σ', p', u' or c'. The quantities to be substituted for the cases "no standard given" and "standard given" are shown below immediately after each relation.

average, $\overline{X}$

$$\sigma_{\overline{X}}' = \frac{\sigma'}{\sqrt{n}} \tag{45}$$

where σ' is the standard deviation of the universe. For no standard given, substitute $\sigma' = \bar{s}/c_4$ or $\sigma' = \overline{R}/d_2$; for standard given, substitute $\sigma' = \sigma_0'$. Equation 44 does not assume a Normal distribution (see Ref *1*, pp. 180 and 181).

standard deviation, s

$$\sigma_s' = (1 - c_4{}^2)^{1/2}\sigma' \tag{46}$$

or

$$= [2(n-1) - 2(n-1)c_4{}^2]^{1/2} \frac{\sigma'}{\sqrt{2(n-1)}} \tag{47}$$

where c_4 is defined in Eq 42, and σ' is the standard deviation of the Normal universe sampled. (Equation 46 is taken from Eq 12, Ref *10*.) For no standard given, substitute $\sigma' = \bar{s}/c_4$ or $\bar{R}/d_2$; for standard given, substitute $\sigma' = \sigma_0{}'$. For control chart purposes, these relations may be used for distributions other than Normal.

approximations to relation for $\sigma_s{}'$

$$\sigma_s{}' = \left(\frac{1}{2(n-1)} - \frac{1}{8n(n-1)} - \frac{3}{16n^2(n-1)} - \cdots\right)^{1/2} \sigma' \tag{48}$$

$$\sigma_s{}' \doteq \left(1 - \frac{1}{8n} - \frac{25}{128n^2}\right) \frac{\sigma'}{\sqrt{2(n-1)}} \tag{49}$$

$$\sigma_s{}' \doteq \frac{\sigma'}{\sqrt{2(n-1)}} \tag{50}$$

where σ' is the standard deviation of the universe [*11*].

The exact relation of Eq 46 or Eq 47 is used in Part 3 for control chart analyses involving $\sigma_s{}'$ and for the determination of factors B_3 and B_4 of Table 2, and B_5 and B_6 of Table 3.

Range R

$$\sigma_R{}' = d_3\sigma' \tag{51}$$

where σ' is the standard deviation of the universe. For no standard given, substitute $\sigma' = \bar{s}/c_4$ or R/d_2; for standard given, substitute $\sigma' = \sigma_0{}'$.

The factor d_3 given in Eq 44 represents the standard deviation for ranges in terms of the true standard deviation of the Normal universe [*11*].

fraction defective, p

$$\sigma_p{}' = \sqrt{\frac{p'(1-p')}{n}} \tag{52}$$

where p' is the value of the fraction defective for the universe. For no standard given, substitute $p' = 3$; for standard given, substitute $p' = p_0{}'$. When p′ is so small that the term $(1 - p')$ may be neglected, the following approximation is used

$$\sigma_p{}' \doteq \sqrt{\frac{p'}{n}} \tag{53}$$

number of defectives, *pn*

$$\sigma'_{pn} = \sqrt{p'n(1 - p')} \tag{54}$$

where p' is the value of the fraction defective for the universe. For no standard given, substitute $p' = \bar{p}$; and for standard given, substitute $p' = p_0'$. The function *pn* has been widely used to represent the number of defectives (defective units) for one or more characteristics.

Both *p* and *pn* have a binomial distribution. Equations 52 and 54 are based on the binomial distribution in which the theoretical frequencies for $pn = 0, 1, 2, \ldots, n$ are given by the first, second, third, etc. terms of the expansion of the binomial $[(1 - p') + p']^n$, where p' is the universe value. When p' is so small that the term $(1 - p')$ may be neglected, the following approximation is used

$$\sigma'_{pn} = \sqrt{p'n} \tag{55}$$

defects per unit, *u*

$$\sigma_u' = \sqrt{\frac{u'}{n}} \tag{56}$$

where *n* is the number of units in sample, and u' is the value of *defects per unit* for the universe. For no standard given, substitute $u' = \bar{u}$; for standard given, substitute $u' = u_0'$.

The number of defects found on any one unit may be considered to result from an unknown but large (practically infinite) number of points where a defect could possibly occur combined with an unknown but very small probability of occurrence at any one point. This leads to the use of the Poisson exponential distribution for which the standard deviation is the square root of the expected number of defects on a single unit. This distribution is likewise applicable to sums of such numbers, such as the observed values of *c*, and to averages of such numbers, such as observed values of *u*, the expected standard deviation of the averages being $1/n$ times that of the sums. Where the number of defects found on any one unit results from a known number of potential causes (relatively a small number as compared with the case described above), and the distribution of the defects per unit is more exactly a multinomial distribution, for practical purposes in most instances the Poisson exponential distribution, although an approximation, may be used for control chart work.

number of defects, *c*

$$\sigma_c' = \sqrt{u'n} = \sqrt{c'} \tag{57}$$

where n is the number of units in sample, u' is the value of *defects per unit* for the universe, and c' is the number of defects in samples of size n for the universe. For no standard given, substitute $c' = \bar{c} = \bar{u}n$; for standard given, substitute $c' = c_0' = u_0'n$. The distribution of the observed values of c is discussed above.

Factors for Computing Control Limits

Note that all these factors are actually functions of n only and the constant 3 resulting from the choice of 3-sigma limits.

averages

$$A = \frac{3}{\sqrt{n}} \tag{58}$$

$$A_3 = \frac{3}{c_4\sqrt{n}} \tag{59}$$

$$A_2 = \frac{3}{d_2\sqrt{n}} \tag{60}$$

NOTE—$A_3 = A/c_4$, $A_2 = A/d_2$.

standard deviations

$$B_5 = c_4 - 3(1 - c_4{}^2)^{1/2} \tag{61}$$

$$B_6 = c_4 + 3(1 - c_4{}^2)^{1/2} \tag{62}$$

$$B_3 = 1 - \frac{3}{c_4}(1 - c_4{}^2)^{1/2} \tag{63}$$

$$B_4 = 1 + \frac{3}{c_4}(1 - c_4{}^2)^{1/2} \tag{64}$$

NOTE—$B_3 = B_5/c_4$, $B_4 = B_6/c_4$.

ranges

$$D_1 = d_2 - 3d_3 \tag{65}$$

$$D_2 = d_2 + 3d_3 \tag{66}$$

$$D_3 = 1 - 3\frac{d_3}{d_2} \tag{67}$$

$$D_4 = 1 + 3\frac{d_3}{d_2} \tag{68}$$

NOTE—$D_3 = D_1/d_2$, $D_4 = D_2/d_2$.

individuals

$$E_3 = \frac{3}{c_4} \tag{69}$$

$$E_2 = \frac{3}{d_2} \tag{70}$$

Supplement B

Explanatory Notes

Note 1—As explained in detail in Supplement A $\sigma_{\bar{X}}'$ and σ_s' are given values computed [*1*] from the observed variation of individual values *within* subgroups and the size n of a subgroup for the first use (A) Control—No Standard Given, and (2) from the adopted standard value of σ' and the size n of a subgroup for the second use (B) Control with Respect to a Given Standard. Likewise, for the first use, σ_p' is given a value computed from the *over-all* value of p, designated $\bar{p}$, and n, and for the second use from p_0' and n. The method for determining σ_R' is outlined in Supplement A.

Note 2—This is discussed fully by Shewhart [*1*]. In some situations in industry in which it is important to catch trouble even if it entails a considerable amount of otherwise unnecessary investigation, 2-sigma limits have been found useful. The necessary changes in the factors for control chart limits will be apparent from their derivation in the text and in Supplement A. Alternatively, in process-quality-control work, probability control limits based on percentage points are sometimes used [*12*].

Note 3—From the viewpoint of the theory of estimation, if normality is assumed, an unbiased and efficient estimate of the standard deviation within subgroups is

$$\frac{1}{c_4}\sqrt{\frac{(n_1 - 1)s_1{}^2 + \cdots + (n_k - 1)s_k{}^2}{n_1 + \cdots + n_k - k}} \tag{71}$$

where c_4 is to be found from Table 2, corresponding to $n = n_1 + \cdots + n_k - k + 1$. Actually, c_4 will be very close to unity if $n_1 + \cdots + n_k - k + 1$ is as large as 15 or more as it usually is, whether n_1, n_2, etc. be large, small, equal, or unequal (see Table 27).

Equations 4, 6, and 9, and the procedure of Sections 8 and 9, "Control—No Standard Given," have been adopted for use in Part 3 with practical considerations in mind, Eq 6 representing a departure from that previously given. From the viewpoint of the theory of estimation they are unbiased or nearly so when used with the appropriate factors as described in the text and are nearly as efficient as Formula 71.

It should be pointed out that the problem of choosing a control chart criterion for use in "Control—No Standard Given" is not essentially a problem in estimation. The criterion is by nature more a test of consistency of the data themselves and must be based on the data at hand including some which may have been influenced by the assignable causes which it is desired to detect. The final justification of a control chart criterion is its proved ability to detect assignable causes economically under practical conditions.

When control has been achieved and standard values are to be based on the observed data, the problem is more a problem in estimation, although in practice many of the assumptions made in estimation theory are imperfectly met and practical considerations, sampling trials, and experience are deciding factors.

In both cases, data are usually plentiful and efficiency of estimation a minor consideration.

Note 4—If most of the samples are of approximately equal size, effort may be saved by first computing and plotting approximate control limits based on some typical sample size, such as the most frequent sample size, a standard sample size, or the average sample size. Then, for any point questionably near the limits, the correct limits based on the actual sample size for that point should be computed and also plotted, if the point would otherwise be shown in incorrect relation to the limits.

Note 5—Here it is of interest to note the nature of the statistical distributions involved, as follows.

(*a*) With respect to a characteristic for which it is possible for only one defect to occur on a unit, and, in general, when the result of examining a unit is to classify it as defective or nondefective by any criterion, the underlying distribution function may often usefully be assumed to be the binomial, where p is the fraction defective and n is the number of units in the sample (for example, see Formula 14 in this part).

(*b*) With respect to a characteristic for which it is possible for two, three, or some other limited number of defects to occur on a unit, such as poor soldered connections on a unit of wired equipment, where we are primarily concerned with the classification of soldered connections, rather than units, into defective and nondefective, the underlying distribution may often usefully be assumed to be the binomial, where p is the ratio of the observed to the possible number of occurrences of defects in the sample and n is the possible number of occurrences of defects in the sample instead of the sample size (for example, see Formula 14 in this part, with n defined as number of possible occurrences per sample).

(*c*) With respect to a characteristic for which it is possible for a large but indeterminate number of defects to occur on a unit, such as finish defects on a painted surface, the underlying distribution may often usefully be assumed to be the Poisson distribution. (The proportion of defects expected in the sample, p, is indeterminate and usually small; and the possible number of occurrences of defects in the sample, n, is also indeterminate and usually large; but the product pn is finite. For the sample this pn value is c.) (For example, see Formula 22 of this part.)

For characteristics of Types (*a*) and (*b*) the fraction p is almost invariably small, say less than 0.10, and under these circumstances the Poisson distribution may be used as a satisfactory approximation to the binomial. Hence, in general, for all these three types of characteristics, taken individually or collectively, we may use relations based on the Poisson distribution. The relations given for control limits for number of defects (Sections 16 and 26) have accordingly been based directly on the Poisson distribution, and the relations for control limits for defects per unit (Sections 15 and 25), have been based indirectly thereon.

Note 6—In the control of a process, it is common practice to extend the central line and control limits on a control chart to cover a future period of operations (see Ref *2*, p. 26). This practice constitutes control with respect to a standard set by previous operating experience and is a simple way to apply this principle when no change in sample size or sizes is contemplated.

When it is not convenient to specify the sample size or sizes in advance, standard values of $\bar{X}'$, σ', etc. may be derived from past control chart data using the relations

$$\bar{X}_0' = \bar{\bar{X}} \qquad\qquad p_0'n = \bar{p}n$$

$$\sigma_0' = \frac{\bar{R}}{d_2} \text{ or } \frac{\bar{s}}{c_4} \qquad\qquad u_0' = \bar{u}$$

$$p_0' = \bar{p} \qquad\qquad c_0' = \bar{c}$$

where the values on the right-hand side of the relations are derived from past data. In this process a certain amount of arbitrary judgment may be used in omitting data from subgroups found or believed to be out of control.

Note 7—It may be of interest to note that, for a given set of data, the mean moving range as defined here is the average of the two values of $\bar{R}$ which would be obtained using ordinary ranges of subgroups of two, starting in one case with the first observation and in the other with the second observation.

The mean moving range is capable of much wider definition [*12*] but that given here has been the one used most in process quality control.

When a control chart for averages and a control chart for ranges are used together, the chart for ranges gives information which is not contained in the chart for averages and the combination is very effective in process control. The combination of a control chart for individuals and a control chart for moving ranges does not possess this dual property, all the information in the chart for moving ranges is contained, somewhat less explicitly, in the chart for individuals.

References

[1] Shewhart, W. A., *Economic Control of Quality of Manufactured Product*, Van Nostrand, New York, 1931.

[2] ANSI Standards Z 1.1-1958(R1975), "Guide for Quality Control;" Z 1.2-1958(R1975), "Control Chart Method of Analyzing Data;" and Z 1.3-1958(R1975), "Control Chart Method of Controlling Quality During Production," American National Standards Institute, New York.

[3] Simon, L. E., *An Engineer's Manual of Statistical Methods*, Wiley, New York, 1941.

[4] British Standard 600:1935, Pearson, E. S., "The Application of Statistical Methods to Industrial Standardization and Quality Control;" British Standard 600 R:1942, Dudding, B. P. and Jennett, W. J., "Quality Coatrol Charts," British Standards Institution, London, England.

[5] Bowker, A. H. and Lieberman, G. L., *Engineering Statistics*, 2nd ed., Prentice-Hall, Englewood Cliffs, N. J., 1972.

[6] Burr, I. W., *Engineering Statistics and Quality Control*, McGraw-Hill, New York, 1953.

[7] Duncan, A. J., *Quality Control and Industrial Statistics*, 4th ed., Irwin, Homewood, Ill., 1974.

[8] Grant, E. L. and Leavenworth, R. S., *Statistical Quality Control*, 4th ed., McGraw-Hill, New York, 1972.

[9] Tippett, L. H. C., "On the Extreme Individuals and the Range of Samples from a Normal Population," *Biometrika*, Vol. 17, 1925.

[10] Mosteller, Frederick, *The Annals of Mathematical Statistics*, Vol. 17, No. 4, Dec. 1946, pp. 377–408.

[11] Romanovsky, U., *Metron*, Vol. 5, No. 4, 1925, pp. 3–46.

[12] Hoel, P. G., *The Annals of Mathematical Statistics*, Vol. 17, No. 4, Dec. 1946, p. 475.

Selected Papers on Other Control Chart Techniques[3]

Barnard, G. A., "Control Charts and Stochastic Processes," *Journal of the Royal Statistical Society*, Series B, Vol. 21, No. 2, 1959, pp. 239–271.

Freund, R. A., "Acceptance Control Charts," *Industrial Quality Control*, Vol. 14, No. 4, Oct. 1957, pp. 13–23.

Freund, R. A., "A Reconsideration of the Variables Control Chart," *Industrial Quality Control*, Vol. 16, No. 11, May 1960, pp. 35–41.

Freund, R. A., "Graphical Process Control," *Industrial Quality Control*, Jan. 1962, pp. 15–22.

[3] Used more for control purposes than presentation of data.

Johnson, N. L. and Leone, F. C., "Comulative Sum Control Charts—Mathematical Principles Applied to Their Construction and Use," *Industrial Quality Control*, June, 1962, pp. 15–21; July 1962, pp. 29–36; and Aug. 1962, pp. 22–28.

Kemp, K. W., "The Average Run Length of the Cumulative Sum Chart When a V-Mask is Used," *Journal of the Royal Statistical Society*, Series B, Vol. 23, 1961, pp. 149–153.

Kemp, K. W., "The Use of Cumulative Sums for Sampling Inspection Schemes," *Applied Statistics*, Vol. 11, 1962, pp. 16–31.

Page, E. S., "Cumulative Sum Charts," *Technometrics*, Vol. 3, No. 1, Feb. 1961, pp. 1–9.

Roberts, S. W., "Control Chart Tests Based on Geometric Moving Averages," *Technometrics*, Vol. 1, No. 3, Aug. 1959, pp. 239–250.

Appendixes

APPENDIX I

TABLES OF SQUARES AND SQUARE ROOTS

SQUARES AND SQUARE ROOTS 1–200

No.	Square	Square Root	No.	Square	Square Root	No.	Square	Square Root	No.	Square	Square Root
1	1	1.0000	51	2 601	7.1414	101	10 201	10.0499	151	22 801	12.2882
2	4	1.4142	52	2 704	7.2111	102	10 404	10.0995	152	23 104	12.3288
3	9	1.7321	53	2 809	7.2801	103	10 609	10.1489	153	23 409	12.3693
4	16	2.0000	54	2 916	7.3485	104	10 816	10.1980	154	23 716	12.4097
5	25	2.2361	55	3 025	7.4162	105	11 025	10.2470	155	24 025	12.4499
6	36	2.4495	56	3 136	7.4833	106	11 236	10.2956	156	24 336	12.4900
7	49	2.6458	57	3 249	7.5498	107	11 449	10.3441	157	24 649	12.5300
8	64	2.8284	58	3 364	7.6158	108	11 664	10.3923	158	24 964	12.5698
9	81	3.0000	59	3 481	7.6811	109	11 881	10.4403	159	25 281	12.6095
10	100	3.1623	60	3 600	7.7460	110	12 100	10.4881	160	25 600	12.6491
11	121	3.3166	61	3 721	7.8102	111	12 321	10.5357	161	25 921	12.6886
12	144	3.4641	62	3 844	7.8740	112	12 544	10.5830	162	26 244	12.7279
13	169	3.6056	63	3 969	7.9373	113	12 769	10.6301	163	26 569	12.7671
14	196	3.7417	64	4 096	8.0000	114	12 996	10.6771	164	26 896	12.8062
15	225	3.8730	65	4 225	8.0623	115	13 225	10.7238	165	27 225	12.8452
16	256	4.0000	66	4 356	8.1240	116	13 456	10.7703	166	27 556	12.8841
17	289	4.1231	67	4 489	8.1854	117	13 689	10.8167	167	27 889	12.9228
18	324	4.2426	68	4 624	8.2462	118	13 924	10.8628	168	28 224	12.9615
19	361	4.3589	69	4 761	8.3066	119	14 161	10.9087	169	28 561	13.0000
20	400	4.4721	70	4 900	8.3666	120	14 400	10.9545	170	28 900	13.0384
21	441	4.5826	71	5 041	8.4261	121	14 641	11.0000	171	29 241	13.0767
22	484	4.6904	72	5 184	8.4853	122	14 884	11.0454	172	29 584	13.1149
23	529	4.7958	73	5 329	8.5440	123	15 129	11.0905	173	29 929	13.1529
24	576	4.8990	74	5 476	8.6023	124	15 376	11.1355	174	30 276	13.1909
25	625	5.0000	75	5 625	8.6603	125	15 625	11.1803	175	30 625	13.2288
26	676	5.0990	76	5 776	8.7178	126	15 876	11.2250	176	30 976	13.2665
27	729	5.1962	77	5 929	8.7750	127	16 129	11.2694	177	31 329	13.3041
28	784	5.2915	78	6 084	8.8318	128	16 384	11.3137	178	31 684	13.3417
29	841	5.3852	79	6 241	8.8882	129	16 641	11.3578	179	32 041	13.3791
30	900	5.4772	80	6 400	8.9443	130	16 900	11.4018	180	32 400	13.4164
31	961	5.5678	81	6 561	9.0000	131	17 161	11.4455	181	32 761	13.4536
32	1 024	5.6569	82	6 724	9.0554	132	17 424	11.4891	182	33 124	13.4907
33	1 089	5.7446	83	6 889	9.1104	133	17 689	11.5326	183	33 489	13.5277
34	1 156	5.8310	84	7 056	9.1652	134	17 956	11.5758	184	33 856	13.5647
35	1 225	5.9161	85	7 225	9.2195	135	18 225	11.6190	185	34 225	13.6015
36	1 296	6.0000	86	7 396	9.2736	136	18 496	11.6619	186	34 596	13.6382
37	1 369	6.0828	87	7 569	9.3274	137	18 769	11.7047	187	34 969	13.6748
38	1 444	6.1644	88	7 744	9.3808	138	19 044	11.7473	188	35 344	13.7113
39	1 521	6.2450	89	7 921	9.4340	139	19 321	11.7898	189	35 721	13.7477
40	1 600	6.3246	90	8 100	9.4868	140	19 600	11.8322	190	36 100	13.7840
41	1 681	6.4031	91	8 281	9.5394	141	19 881	11.8743	191	36 481	13.8203
42	1 764	6.4807	92	8 464	9.5917	142	20 164	11.9164	192	36 864	13.8564
43	1 849	6.5574	93	8 649	9.6437	143	20 449	11.9583	193	37 249	13.8924
44	1 936	6.6332	94	8 836	9.6954	144	20 736	12.0000	194	37 636	13.9284
45	2 025	6.7082	95	9 025	9.7468	145	21 025	12.0416	195	38 025	13.9642
46	2 116	6.7823	96	9 216	9.7980	146	21 316	12.0830	196	38 416	14.0000
47	2 209	6.8557	97	9 409	9.8489	147	21 609	12.1244	197	38 809	14.0357
48	2 304	6.9282	98	9 604	9.8995	148	21 904	12.1655	198	39 204	14.0712
49	2 401	7.0000	99	9 801	9.9499	149	22 201	12.2066	199	39 [illegible]	14.1067
50	2 500	7.0711	100	10 000	10.0000	150	22 500	12.2474	200	40 000	14.1421

NOTE.—To find the square root of numbers greater than 2000, look up the given number in the "Square" column and obtain the answer from the corresponding value in the "No." column

Example 1.—To find the square root of 2174.386

Group the digits in pairs starting from the decimal point, thus:

$$\overset{\frown}{21}\ \overset{\frown}{74}.\ \overset{\frown}{38}\ \overset{\frown}{60}$$

There will always be one digit in the square root for each group in the given number. Observe that the square root of the number in the first group (21) is between 4 and 5. Referring to the table, find in the "No." column the two numbers lying between 400 and 500 whose squares most nearly equal the given number. In this case the numbers are 466 and 467. Interpolating and locating the decimal point gives 46.63 for the desired square root.

In referring to the table, numbers between 40 and 50 in the "No." column may be used as well as those between 400 and 500, but the latter numbers will give the desired square root to one more significant figure.

Example 2.—To find the square root of 21743.86.

Group the digits in pairs as before:

$$\overset{\frown}{2}\ \overset{\frown}{17}\ \overset{\frown}{43}.\ \overset{\frown}{86}$$

and observe that the square root of the number in the first group (2) is between 1 and 2. Refer to the "No." column of the table for numbers lying between 1000 and 2000. In this case, the numbers are 1474 and 1475, and interpolating and locating the decimal point gives 147.46.

201–500 SQUARES AND SQUARE ROOTS (*Continued*)

No	Square	Square Root	No.	Square	Square Root	No.	Square	Square Root	No.	Square	Square Root
201	40 401	14.1774	276	76 176	16.6132	351	123 201	18.7350	426	181 476	20.6398
202	40 804	14.2127	277	76 729	16.6433	352	123 904	18.7617	427	182 329	20.6640
203	41 209	14.2478	278	77 284	16.6733	353	124 609	18.7883	428	183 184	20.6882
204	41 616	14.2829	279	77 841	16.7033	354	125 316	18.8149	429	184 041	20.7123
205	42 025	14.3178	280	78 400	16.7332	355	126 025	18.8414	430	184 900	20.7364
206	42 436	14.3527	281	78 961	16.7631	356	126 736	18.8680	431	185 761	20.7605
207	42 849	14.3875	282	79 524	16.7929	357	127 449	18.8944	432	186 624	20.7846
208	43 264	14.4222	283	80 089	16.8226	358	128 164	18.9209	433	187 489	20.8087
209	43 681	14.4568	284	80 656	16.8523	359	128 881	18.9473	434	188 356	20.8327
210	44 100	14.4914	285	81 225	16.8819	360	129 600	18.9737	435	189 225	20.8567
211	44 521	14.5258	286	81 796	16.9115	361	130 321	19.0000	436	190 096	20.8806
212	44 944	14.5602	287	82 369	16.9411	362	131 044	19.0263	437	190 969	20.9045
213	45 369	14.5945	288	82 944	16.9706	363	131 769	19.0526	438	191 844	20.9284
214	45 796	14.6287	289	83 521	17.0000	364	132 496	19.0788	439	192 721	20.9523
215	46 225	14.6629	290	84 100	17.0294	365	133 225	19.1050	440	193 600	20.9762
216	46 656	14.6969	291	84 681	17.0587	366	133 956	19.1311	441	194 481	21.0000
217	47 089	14.7309	292	85 264	17.0880	367	134 689	19.1572	442	195 364	21.0238
218	47 524	14.7648	293	85 849	17.1172	368	135 424	19.1833	443	196 249	21.0476
219	47 961	14.7986	294	86 436	17.1464	369	136 161	19.2094	444	197 136	21.0713
220	48 400	14.8324	295	87 025	17.1756	370	136 900	19.2354	445	198 025	21.0950
221	48 841	14.8661	296	87 616	17.2047	371	137 641	19.2614	446	198 916	21.1187
222	49 284	14.8997	297	88 209	17.2337	372	138 384	19.2873	447	199 809	21.1424
223	49 729	14.9332	298	88 804	17.2627	373	139 129	19.3132	448	200 704	21.1660
224	50 176	14.9666	299	89 401	17.2916	374	139 876	19.3391	449	201 601	21.1896
225	50 625	15.0000	300	90 000	17.3205	375	140 625	19.3649	450	202 500	21.2132
226	51 076	15.0333	301	90 601	17.3494	376	141 376	19.3907	451	203 401	21.2368
227	51 529	15.0665	302	91 204	17.3781	377	142 129	19.4165	452	204 304	21.2603
228	51 984	15.0997	303	91 809	17.4069	378	142 884	19.4422	453	205 209	21.2838
229	52 441	15.1327	304	92 416	17.4356	379	143 641	19.4679	454	206 116	21.3073
230	52 900	15.1658	305	93 025	17.4642	380	144 400	19.4936	455	207 025	21.3307
231	53 361	15.1987	306	93 636	17.4929	381	145 161	19.5192	456	207 936	21.3542
232	53 824	15.2315	307	94 249	17.5214	382	145 924	19.5448	457	208 849	21.3776
233	54 289	15.2643	308	94 864	17.5499	383	146 689	19.5704	458	209 764	21.4009
234	54 756	15.2971	309	95 481	17.5784	384	147 456	19.5959	459	210 681	21.4243
235	55 225	15.3297	310	96 100	17.6068	385	148 225	19.6214	460	211 600	21.4476
236	55 696	15.3623	311	96 721	17.6352	386	148 996	19.6469	461	212 521	21.4709
237	56 169	15.3948	312	97 344	17.6635	387	149 769	19.6723	462	213 444	21.4942
238	56 644	15.4272	313	97 969	17.6918	388	150 544	19.6977	463	214 369	21.5174
239	57 121	15.4596	314	98 596	17.7200	389	151 321	19.7231	464	215 296	21.5407
240	57 600	15.4919	315	99 225	17.7482	390	152 100	19.7484	465	216 225	21.5639
241	58 081	15.5242	316	99 856	17.7764	391	152 881	19.7737	466	217 156	21.5870
242	58 564	15.5563	317	100 489	17.8045	392	153 664	19.7990	467	218 089	21.6102
243	59 049	15.5885	318	101 124	17.8326	393	154 449	19.8242	468	219 024	21.6333
244	59 536	15.6205	319	101 761	17.8606	394	155 236	19.8494	469	219 961	21.6564
245	60 025	15.6525	320	102 400	17.8885	395	156 025	19.8746	470	220 900	21.6795
246	60 516	15.6844	321	103 041	17.9165	396	156 816	19.8997	471	221 841	21.7025
247	61 009	15.7162	322	103 684	17.9444	397	157 609	19.9249	472	222 784	21.7256
248	61 504	15.7480	323	104 329	17.9722	398	158 404	19.9499	473	223 729	21.7486
249	62 001	15.7797	324	104 976	18.0000	399	159 201	19.9750	474	224 676	21.7715
250	62 500	15.8114	325	105 625	18.0278	400	160 000	20.0000	475	225 625	21.7945
251	63 001	15.8430	326	106 276	18.0555	401	160 801	20.0250	476	226 576	21.8174
252	63 504	15.8745	327	106 929	18.0831	402	161 604	20.0499	477	227 529	21.8403
253	64 009	15.9060	328	107 584	18.1108	403	162 409	20.0749	478	228 484	21.8632
254	64 516	15.9374	329	108 241	18.1384	404	163 216	20.0998	479	229 441	21.8861
255	65 025	15.9687	330	108 900	18.1659	405	164 025	20.1246	480	230 400	21.9089
256	65 536	16.0000	331	109 561	18.1934	406	164 836	20.1494	481	231 361	21.9317
257	66 049	16.0312	332	110 224	18.2209	407	165 649	20.1742	482	232 324	21.9545
258	66 564	16.0624	333	110 889	18.2483	408	166 464	20.1990	483	233 289	21.9773
259	67 081	16.0935	334	111 556	18.2757	409	167 281	20.2237	484	234 256	22.0000
260	67 600	16.1245	335	112 225	18.3030	410	168 100	20.2485	485	235 225	22.0227
261	68 121	16.1555	336	112 896	18.3303	411	168 921	20.2731	486	236 196	22.0454
262	68 644	16.1864	337	113 569	18.3576	412	169 744	20.2978	487	237 169	22.0681
263	69 169	16.2173	338	114 244	18.3848	413	170 569	20.3224	488	238 144	22.0907
264	69 696	16.2481	339	114 921	18.4120	414	171 396	20.3470	489	239 121	22.1133
265	70 225	16.2788	340	115 600	18.4391	415	172 225	20.3715	490	240 100	22.1359
266	70 756	16.3095	341	116 281	18.4662	416	173 056	20.3961	491	241 081	22.1585
267	71 289	16.3401	342	116 964	18.4932	417	173 889	20.4206	492	242 064	22.1811
268	71 824	16.3707	343	117 649	18.5203	418	174 724	20.4450	493	243 049	22.2036
269	72 361	16.4012	344	118 336	18.5472	419	175 561	20.4695	494	244 036	22.2261
270	72 900	16.4317	345	119 025	18.5742	420	176 400	20.4939	495	245 025	22.2486
271	73 441	16.4621	346	119 716	18.6011	421	177 241	20.5183	496	246 016	22.2711
272	73 984	16.4924	347	120 409	18.6279	422	178 084	20.5426	497	247 009	22.2935
273	74 529	16.5227	348	121 104	18.6548	423	178 929	20.5670	498	248 004	22.3159
274	75 076	16.5529	349	121 801	18.6815	424	179 776	20.5913	499	249 001	22.3383
275	75 625	16.5831	350	122 500	18.7083	425	180 625	20.6155	500	250 000	22.3607

SQUARES AND SQUARE ROOTS (*Continued*) **501–800**

No.	Square	Square Root	No.	Square	Square Root	No.	Square	Square Root	No.	Square	Square Root
501	251 001	22.3830	576	331 776	24.0000	651	423 801	25.5147	726	527 076	26.9444
502	252 004	22.4054	577	332 929	24.0208	652	425 104	25.5343	727	528 529	26.9629
503	253 009	22.4277	578	334 084	24.0416	653	426 409	25.5539	728	529 984	26.9815
504	254 016	22.4499	579	335 241	24.0624	654	427 716	25.5734	729	531 441	27.0000
505	255 025	22.4722	580	336 400	24.0832	655	429 025	25.5930	730	532 900	27.0185
506	256 036	22.4944	581	337 561	24.1039	656	430 336	25.6125	731	534 361	27.0370
507	257 049	22.5167	582	338 724	24.1247	657	431 649	25.6320	732	535 824	27.0555
508	258 064	22.5389	583	339 889	24.1454	658	432 964	25.6515	733	537 289	27.0740
509	259 081	22.5610	584	341 056	24.1661	659	434 281	25.6710	734	538 756	27.0924
510	260 100	22.5832	585	342 225	24.1868	660	435 600	25.6905	735	540 225	27.1109
511	261 121	22.6053	586	343 396	24.2074	661	436 921	25.7099	736	541 696	27.1293
512	262 144	22.6274	587	344 569	24.2281	662	438 244	25.7294	737	543 169	27.1477
513	263 169	22.6495	588	345 744	24.2487	663	439 569	25.7488	738	544 644	27.1662
514	264 196	22.6716	589	346 921	24.2693	664	440 896	25.7682	739	546 121	27.1846
515	265 225	22.6936	590	348 100	24.2899	665	442 225	25.7876	740	547 600	27.2029
516	266 256	22.7156	591	349 281	24.3105	666	443 556	25.8070	741	549 081	27.2213
517	267 289	22.7376	592	350 464	24.3311	667	444 889	25.8263	742	550 564	27.2397
518	268 324	22.7596	593	351 649	24.3516	668	446 224	25.8457	743	552 049	27.2580
519	269 361	22.7816	594	352 836	24.3721	669	447 561	25.8650	744	553 536	27.2764
520	270 400	22.8035	595	354 025	24.3926	670	448 900	25.8844	745	555 025	27.2947
521	271 441	22.8254	596	355 216	24.4131	671	450 241	25.9037	746	556 516	27.3130
522	272 484	22.8473	597	356 409	24.4336	672	451 584	25.9230	747	558 009	27.3313
523	273 529	22.8692	598	357 604	24.4540	673	452 929	25.9422	748	559 504	27.3496
524	274 576	22.8910	599	358 801	24.4745	674	454 276	25.9615	749	561 001	27.3679
525	275 625	22.9129	600	360 000	24.4949	675	455 625	25.9808	750	562 500	27.3861
526	276 676	22.9347	601	361 201	24.5153	676	456 976	26.0000	751	564 001	27.4044
527	277 729	22.9565	602	362 404	24.5357	677	458 329	26.0192	752	565 504	27.4226
528	278 784	22.9783	603	363 609	24.5561	678	459 684	26.0384	753	567 009	27.4408
529	279 841	23.0000	604	364 816	24.5764	679	461 041	26.0576	754	568 516	27.4591
530	280 900	23.0217	605	366 025	24.5967	680	462 400	26.0768	755	570 025	27.4773
531	281 961	23.0434	606	367 236	24.6171	681	463 761	26.0960	756	571 536	27.4955
532	283 024	23.0651	607	368 449	24.6374	682	465 124	26.1151	757	573 049	27.5136
533	284 089	23.0868	608	369 664	24.6577	683	466 489	26.1343	758	574 564	27.5318
534	285 156	23.1084	609	370 881	24.6779	684	467 856	26.1534	759	576 081	27.5500
535	286 225	23.1301	610	372 100	24.6982	685	469 225	26.1725	760	577 600	27.5681
536	287 296	23.1517	611	373 321	24.7184	686	470 596	26.1916	761	579 121	27.5862
537	288 369	23.1733	612	374 544	24.7386	687	471 969	26.2107	762	580 644	27.6043
538	289 444	23.1948	613	375 769	24.7588	688	473 344	26.2298	763	582 169	27.6225
539	290 521	23.2164	614	376 996	24.7790	689	474 721	26.2488	764	583 696	27.6405
540	291 600	23.2379	615	378 225	24.7992	690	476 100	26.2679	765	585 225	27.6586
541	292 681	23.2594	616	379 456	24.8193	691	477 481	26.2869	766	586 756	27.6767
542	293 764	23.2809	617	380 689	24.8395	692	478 864	26.3059	767	588 289	27.6948
543	294 849	23.3024	618	381 924	24.8596	693	480 249	26.3249	768	589 824	27.7128
544	295 936	23.3238	619	383 161	24.8797	694	481 636	26.3439	769	591 361	27.7308
545	297 025	23.3452	620	384 400	24.8998	695	483 025	26.3629	770	592 900	27.7489
546	298 116	23.3666	621	385 641	24.9199	696	484 416	26.3818	771	594 441	27.7669
547	299 209	23.3880	622	386 884	24.9399	697	485 809	26.4008	772	595 984	27.7849
548	300 304	23.4094	623	388 129	24.9600	698	487 204	26.4197	773	597 529	27.8029
549	301 401	23.4307	624	389 376	24.9800	699	488 601	26.4386	774	599 076	27.8209
550	302 500	23.4521	625	390 625	25.0000	700	490 000	26.4575	775	600 625	27.8388
551	303 601	23.4734	626	391 876	25.0200	701	491 401	26.4764	776	602 176	27.8568
552	304 704	23.4947	627	393 129	25.0400	702	492 804	26.4953	777	603 729	27.8747
553	305 809	23.5160	628	394 384	25.0599	703	494 209	26.5141	778	605 284	27.8927
554	306 916	23.5372	629	395 641	25.0799	704	495 616	26.5330	779	606 841	27.9106
555	308 025	23.5584	630	396 900	25.0998	705	497 025	26.5518	780	608 400	27.9285
556	309 136	23.5797	631	398 161	25.1197	706	498 436	26.5707	781	609 961	27.9464
557	310 249	23.6008	632	399 424	25.1396	707	499 849	26.5895	782	611 524	27.9643
558	311 364	23.6220	633	400 689	25.1595	708	501 264	26.6083	783	613 089	27.9821
559	312 481	23.6432	634	401 956	25.1794	709	502 681	26.6271	784	614 656	28.0000
560	313 600	23.6643	635	403 225	25.1992	710	504 100	26.6458	785	616 225	28.0179
561	314 721	23.6854	636	404 496	25.2190	711	505 521	26.6646	786	617 796	28.0357
562	315 844	23.7065	637	405 769	25.2389	712	506 944	26.6833	787	619 369	28.0535
563	316 969	23.7276	638	407 044	25.2587	713	508 369	26.7021	788	620 944	28.0713
564	318 096	23.7487	639	408 321	25.2784	714	509 796	26.7208	789	622 521	28.0891
565	319 225	23.7697	640	409 600	25.2982	715	511 225	26.7395	790	624 100	28.1069
566	320 356	23.7908	641	410 881	25.3180	716	512 656	26.7582	791	625 681	28.1247
567	321 489	23.8118	642	412 164	25.3377	717	514 089	26.7769	792	627 264	28.1425
568	322 624	23.8328	643	413 449	25.3574	718	515 524	26.7955	793	628 849	28.1603
569	323 761	23.8537	644	414 736	25.3772	719	516 961	26.8142	794	630 436	28.1780
570	324 900	23.8747	645	416 025	25.3969	720	518 400	26.8328	795	632 025	28.1957
571	326 041	23.8956	646	417 316	25.4165	721	519 841	26.8514	796	633 616	28.2135
572	327 184	23.9165	647	418 609	25.4362	722	521 284	26.8701	797	635 209	28.2312
573	328 329	23.9374	648	419 904	25.4558	723	522 729	26.8887	798	636 804	28.2489
574	329 476	23.9583	649	421 201	25.4755	724	524 176	26.9072	799	638 401	28.2666
575	330 625	23.9792	650	422 500	25.4951	725	525 625	26.9258	800	640 000	28.2843

801–1100 SQUARES AND SQUARE ROOTS (*Continued*)

No.	Square	Square Root	No.	Square	Square Root	No.	Square	Square Root	No.	Square	Square Root
801	641 601	28.3019	876	767 376	29.5973	951	904 401	30.8383	1026	1 052 676	32.0312
802	643 204	28.3196	877	769 129	29.6142	952	906 304	30.8545	1027	1 054 729	32.0468
803	644 809	28.3373	878	770 884	29.6311	953	908 209	30.8707	1028	1 056 784	32.0624
804	646 416	28.3549	879	772 641	29.6479	954	910 116	30.8869	1029	1 058 841	32.0780
805	648 025	28.3725	880	774 400	29.6648	955	912 025	30.9031	1030	1 060 900	32.0936
806	649 636	28.3901	881	776 161	29.6816	956	913 936	30.9192	1031	1 062 961	32.1092
807	651 249	28.4077	882	777 924	29.6985	957	915 849	30.9354	1032	1 065 024	32.1248
808	652 864	28.4253	883	779 689	29.7153	958	917 764	30.9516	1033	1 067 089	32.1403
809	654 481	28.4429	884	781 456	29.7321	959	919 681	30.9677	1034	1 069 156	32.1559
810	656 100	28.4605	885	783 225	29.7489	960	921 600	30.9839	1035	1 071 225	32.1714
811	657 721	28.4781	886	784 996	29.7658	961	923 521	31.0000	1036	1 073 296	32.1870
812	659 344	28.4956	887	786 769	29.7825	962	925 444	31.0161	1037	1 075 369	32.2025
813	660 969	28.5132	888	788 544	29.7993	963	927 369	31.0322	1038	1 077 444	32.2180
814	662 596	28.5307	889	790 321	29.8161	964	929 296	31.0483	1039	1 079 521	32.2335
815	664 225	28.5482	890	792 100	29.8329	965	931 225	31.0644	1040	1 081 600	32.2490
816	665 856	28.5657	891	793 881	29.8496	966	933 156	31.0805	1041	1 083 681	32.2645
817	667 489	28.5832	892	795 664	29.8664	967	935 089	31.0966	1042	1 085 764	32.2800
818	669 124	28.6007	893	797 449	29.8831	968	937 024	31.1127	1043	1 087 849	32.2955
819	670 761	28.6182	894	799 236	29.8998	969	938 961	31.1288	1044	1 089 936	32.3110
820	672 400	28.6356	895	801 025	29.9166	970	940 900	31.1448	1045	1 092 025	32.3265
821	674 041	28.6531	896	802 816	29.9333	971	942 841	31.1609	1046	1 094 116	32 3419
822	675 684	28.6705	897	804 609	29.9500	972	944 784	31.1769	1047	1 096 209	32.3574
823	677 329	28.6880	898	806 404	29.9666	973	946 729	31.1929	1048	1 098 304	32.3728
824	678 976	28.7054	899	808 201	29.9833	974	948 676	31.2090	1049	1 100 401	32.3883
825	680 625	28.7228	900	810 000	30.0000	975	950 625	31.2250	1050	1 102 500	32.4037
826	682 276	28.7402	901	811 801	30.0167	976	952 576	31.2410	1051	1 104 601	32.4191
827	683 929	28.7576	902	813 604	30.0333	977	954 529	31.2570	1052	1 106 704	32.4345
828	685 584	28.7750	903	815 409	30.0500	978	956 484	31.2730	1053	1 108 809	32.4500
829	687 241	28.7924	904	817 216	30.0666	979	958 441	31.2890	1054	1 110 916	32.4654
830	688 900	28.8097	905	819 025	30.0832	980	960 400	31.3050	1055	1 113 025	32.4808
831	690 561	28 8271	906	820 836	30.0998	981	962 361	31.3209	1056	1 115 136	32.4962
832	692 224	28 8444	907	822 649	30.1164	982	964 324	31.3369	1057	1 117 249	32.5115
833	693 889	28 8617	908	824 464	30.1330	983	966 289	31.3528	1058	1 119 364	32.5269
834	695 556	28 8791	909	826 281	30.1496	984	968 256	31.3688	1059	1 121 481	32.5423
835	697 225	28 8964	910	828 100	30.1662	985	970 225	31.3847	1060	1 123 600	32.5576
836	698 896	28.9137	911	829 921	30.1828	986	972 196	31.4006	1061	1 125 721	32.5730
837	700 569	28.9310	912	831 744	30.1993	987	974 169	31.4166	1062	1 127 844	32.5883
838	702 244	28.9482	913	833 569	30.2159	988	976 144	31.4325	1063	1 129 969	32.6037
839	703 921	28.9655	914	835 396	30.2324	989	978 121	31.4484	1064	1 132 096	32.6190
840	705 600	28.9828	915	837 225	30.2490	990	980 100	31.4643	1065	1 134 225	32.6343
841	707 281	29.0000	916	839 056	30.2655	991	982 081	31.4802	1066	1 136 356	32.6497
842	708 964	29.0172	917	840 889	30.2820	992	984 064	31.4960	1067	1 138 489	32 6650
843	710 649	29.0345	918	842 724	30.2985	993	986 049	31.5119	1068	1 140 624	32.6803
844	712 336	29.0517	919	844 561	30.3150	994	988 036	31.5278	1069	1 142 761	32.6956
845	714 025	29.0689	920	846 400	30.3315	995	990 025	31.5436	1070	1 144 900	32.7109
846	715 716	29.0861	921	848 241	30.3480	996	992 016	31.5595	1071	1 147 041	32.7261
847	717 409	29.1033	922	850 084	30.3645	997	994 009	31.5753	1072	1 149 184	32.7414
848	719 104	29.1204	923	851 929	30.3809	998	996 004	31.5911	1073	1 151 329	32.7567
849	720 801	29.1376	924	853 776	30.3974	999	998 001	31.6070	1074	1 153 476	32.7719
850	722 500	29.1548	925	855 625	30.4138	1000	1 000 000	31.6228	1075	1 155 625	32.7872
851	724 201	29.1719	926	857 476	30.4302	1001	1 002 001	31.6386	1076	1 157 776	32.8024
852	725 904	29.1890	927	859 329	30.4467	1002	1 004 004	31.6544	1077	1 159 929	32.8177
853	727 609	29.2062	928	861 184	30.4631	1003	1 006 009	31.6702	1078	1 162 084	32.8329
854	729 316	29.2233	929	863 041	30.4795	1004	1 008 016	31.6860	1079	1 164 241	32.8481
855	731 025	29.2404	930	864 900	30.4959	1005	1 010 025	31.7017	1080	1 166 400	32.8634
856	732 736	29.2575	931	866 761	30.5123	1006	1 012 036	31.7175	1081	1 168 561	32.8786
857	734 449	29.2746	932	868 624	30.5287	1007	1 014 049	31.7333	1082	1 170 724	32.8938
858	736 164	29.2916	933	870 489	30.5450	1008	1 016 064	31.7490	1083	1 172 889	32.9090
859	737 881	29.3087	934	872 356	30.5614	1009	1 018 081	31.7648	1084	1 175 056	32.9242
860	739 600	29.3258	935	874 225	30.5778	1010	1 020 100	31.7805	1085	1 177 225	32.9393
861	741 321	29.3428	936	876 096	30.5941	1011	1 022 121	31.7962	1086	1 179 396	32.9545
862	743 044	29.3598	937	877 969	30.6105	1012	1 024 144	31.8119	1087	1 181 569	32.9697
863	744 769	29.3769	938	879 844	30.6268	1013	1 026 169	31.8277	1088	1 183 744	32.9848
864	746 496	29.3939	939	881 721	30.6431	1014	1 028 196	31.8434	1089	1 185 921	33.0000
865	748 225	29.4109	940	883 600	30.6594	1015	1 030 225	31.8591	1090	1 188 100	33.0151
866	749 956	29.4279	941	885 481	30.6757	1016	1 032 256	31.8748	1091	1 190 281	33.0303
867	751 689	29.4449	942	887 364	30.6920	1017	1 034 289	31.8904	1092	1 192 464	33.0454
868	753 424	29.4618	943	889 249	30.7083	1018	1 036 324	31.9061	1093	1 194 649	33.0606
869	755 161	29.4788	944	891 136	30.7246	1019	1 038 361	31.9218	1094	1 196 836	33.0757
870	756 900	29.4958	945	893 025	30.7409	1020	1 040 400	31.9374	1095	1 199 025	33.0908
871	758 641	29.5127	946	894 916	30.7571	1021	1 042 441	31.9531	1096	1 201 216	33.1059
872	760 384	29.5296	947	896 809	30.7734	1022	1 044 484	31.9687	1097	1 203 409	33.1210
873	762 129	29.5466	948	898 704	30.7896	1023	1 046 529	31.9844	1098	1 205 604	33.1361
874	763 876	29.5635	949	900 601	30.8058	1024	1 048 576	32.0000	1099	1 207 801	33.1512
875	765 625	29.5804	950	902 500	30.8221	1025	1 050 625	32.0156	1100	1 210 000	33.1662

SQUARES AND SQUARE ROOTS (*Continued*) 1101–1400

No.	Square	Square Root	No.	Square	Square Root	No.	Square	Square Root	No.	Square	Square Root
1101	1 212 201	33.1813	1176	1 382 976	34.2929	1251	1 565 001	35.3695	1326	1 758 276	36.4143
1102	1 214 404	33.1964	1177	1 385 329	34.3074	1252	1 567 504	35.3836	1327	1 760 929	36.4280
1103	1 216 609	33.2114	1178	1 387 684	34.3220	1253	1 570 009	35.3977	1328	1 763 584	36.4417
1104	1 218 816	33.2265	1179	1 390 041	34.3366	1254	1 572 516	35.4119	1329	1 766 241	36.4555
1105	1 221 025	33.2415	1180	1 392 400	34.3511	1255	1 575 025	35.4260	1330	1 768 900	36.4692
1106	1 223 236	33.2566	1181	1 394 761	34.3657	1256	1 577 536	35.4401	1331	1 771 561	36.4829
1107	1 225 449	33.2716	1182	1 397 124	34.3802	1257	1 580 049	35.4542	1332	1 774 224	36.4966
1108	1 227 664	33.2866	1183	1 399 489	34.3948	1258	1 582 564	35.4683	1333	1 776 889	36.5103
1109	1 229 881	33.3017	1184	1 401 856	34.4093	1259	1 585 081	35.4824	1334	1 779 556	36.5240
1110	1 232 100	33.3167	1185	1 404 225	34.4238	1260	1 587 600	35.4965	1335	1 782 225	36.5377
1111	1 234 321	33.3317	1186	1 406 596	34.4384	1261	1 590 121	35.5106	1336	1 784 896	36.5513
1112	1 236 544	33.3467	1187	1 408 969	34.4529	1262	1 592 644	35.5246	1337	1 787 569	36.5650
1113	1 238 769	33.3617	1188	1 411 344	34.4674	1263	1 595 169	35.5387	1338	1 790 244	36.5787
1114	1 240 996	33.3766	1189	1 413 721	34.4819	1264	1 597 696	35.5528	1339	1 792 921	36.5923
1115	1 243 225	33.3916	1190	1 416 100	34.4964	1265	1 600 225	35.5668	1340	1 795 600	36.6060
1116	1 245 456	33.4066	1191	1 418 481	34.5109	1266	1 602 756	35.5809	1341	1 798 281	36.6197
1117	1 247 689	33.4215	1192	1 420 864	34.5254	1267	1 605 289	35.5949	1342	1 800 964	36.6333
1118	1 249 924	33.4365	1193	1 423 249	34.5398	1268	1 607 824	35.6090	1343	1 803 649	36.6470
1119	1 252 161	33.4515	1194	1 425 636	34.5543	1269	1 610 361	35.6230	1344	1 806 336	36.6606
1120	1 254 400	33.4664	1195	1 428 025	34.5688	1270	1 612 900	35.6371	1345	1 809 025	36.6742
1121	1 256 641	33.4813	1196	1 430 416	34.5832	1271	1 615 441	35.6511	1346	1 811 716	36.6879
1122	1 258 884	33.4963	1197	1 432 809	34.5977	1272	1 617 984	35.6651	1347	1 814 409	36.7015
1123	1 261 129	33.5112	1198	1 435 204	34.6121	1273	1 620 529	35.6791	1348	1 817 104	36.7151
1124	1 263 376	33.5261	1199	1 437 601	34.6266	1274	1 623 076	35.6931	1349	1 819 801	36.7287
1125	1 265 625	33.5410	1200	1 440 000	34.6410	1275	1 625 625	35.7071	1350	1 822 500	36.7423
1126	1 267 876	33.5559	1201	1 442 401	34.6554	1276	1 628 176	35.7211	1351	1 825 201	36.7560
1127	1 270 129	33.5708	1202	1 444 804	34.6699	1277	1 630 729	35.7351	1352	1 827 904	36.7696
1128	1 272 384	33.5857	1203	1 447 209	34.6843	1278	1 633 284	35.7491	1353	1 830 609	36.7831
1129	1 274 641	33.6006	1204	1 449 616	34.6987	1279	1 635 841	35.7631	1354	1 833 316	36.7967
1130	1 276 900	33.6155	1205	1 452 025	34.7131	1280	1 638 400	35.7771	1355	1 836 025	36.8103
1131	1 279 161	33.6303	1206	1 454 436	34.7275	1281	1 640 961	35.7911	1356	1 838 736	36.8239
1132	1 281 424	33.6452	1207	1 456 849	34.7419	1282	1 643 524	35.8050	1357	1 841 449	36.8375
1133	1 283 689	33.6601	1208	1 459 264	34.7563	1283	1 646 089	35.8190	1358	1 844 164	36.8511
1134	1 285 956	33.6749	1209	1 461 681	34.7707	1284	1 648 656	35.8329	1359	1 846 881	36.8646
1135	1 288 225	33.6898	1210	1 464 100	34.7851	1285	1 651 225	35.8469	1360	1 849 600	36.8782
1136	1 290 496	33.7046	1211	1 466 521	34.7994	1286	1 653 796	35.8608	1361	1 852 321	36.8917
1137	1 292 769	33.7194	1212	1 468 944	34.8138	1287	1 656 369	35.8748	1362	1 855 044	36.9053
1138	1 295 044	33.7343	1213	1 471 369	34.8281	1288	1 658 944	35.8887	1363	1 857 769	36.9188
1139	1 297 321	33.7491	1214	1 473 796	34.8425	1289	1 661 521	35.9026	1364	1 860 496	36.9324
1140	1 299 600	33.7639	1215	1 476 225	34.8569	1290	1 664 100	35.9166	1365	1 863 225	36.9459
1141	1 301 881	33.7787	1216	1 478 656	34.8712	1291	1 666 681	35.9305	1366	1 865 956	36.9594
1142	1 304 164	33.7935	1217	1 481 089	34.8855	1292	1 669 264	35.9444	1367	1 868 689	36.9730
1143	1 306 449	33.8083	1218	1 483 524	34.8999	1293	1 671 849	35.9583	1368	1 871 424	36.9865
1144	1 308 736	33.8231	1219	1 485 961	34.9142	1294	1 674 436	35.9722	1369	1 874 161	37.0000
1145	1 311 025	33.8378	1220	1 488 400	34.9285	1295	1 677 025	35.9861	1370	1 876 900	37.0135
1146	1 313 316	33.8526	1221	1 490 841	34.9428	1296	1 679 616	36.0000	1371	1 879 641	37.0270
1147	1 315 609	33.8674	1222	1 493 284	34.9571	1297	1 682 209	36.0139	1372	1 882 384	37.0405
1148	1 317 904	33.8821	1223	1 495 729	34.9714	1298	1 684 804	36.0278	1373	1 885 129	37.0540
1149	1 320 201	33.8969	1224	1 498 176	34.9857	1299	1 687 401	36.0416	1374	1 887 876	37.0675
1150	1 322 500	33.9116	1225	1 500 625	35.0000	1300	1 690 000	36.0555	1375	1 890 625	37.0810
1151	1 324 801	33.9264	1226	1 503 076	35.0143	1301	1 692 601	36.0694	1376	1 893 376	37.0945
1152	1 327 104	33.9411	1227	1 505 529	35.0286	1302	1 695 204	36.0832	1377	1 896 129	37.1080
1153	1 329 409	33.9559	1228	1 507 984	35.0428	1303	1 697 809	36.0971	1378	1 898 884	37.1214
1154	1 331 716	33.9706	1229	1 510 441	35.0571	1304	1 700 416	36.1109	1379	1 901 641	37.1349
1155	1 334 025	33.9853	1230	1 512 900	35.0714	1305	1 703 025	36.1248	1380	1 904 400	37.1484
1156	1 336 336	34.0000	1231	1 515 361	35.0856	1306	1 705 636	36.1386	1381	1 907 161	37.1618
1157	1 338 649	34.0147	1232	1 517 824	35.0999	1307	1 708 249	36.1525	1382	1 909 924	37.1753
1158	1 340 964	34.0294	1233	1 520 289	35.1141	1308	1 710 864	36.1663	1383	1 912 689	37.1887
1159	1 343 281	34.0441	1234	1 522 756	35.1283	1309	1 713 481	36.1801	1384	1 915 456	37.2022
1160	1 345 600	34.0588	1235	1 525 225	35.1426	1310	1 716 100	36.1939	1385	1 918 225	37.2156
1161	1 347 921	34.0735	1236	1 527 696	35.1568	1311	1 718 721	36.2077	1386	1 920 996	37.2290
1162	1 350 244	34.0881	1237	1 530 169	35.1710	1312	1 721 344	36.2215	1387	1 923 769	37.2424
1163	1 352 569	34.1028	1238	1 532 644	35.1852	1313	1 723 969	36.2353	1388	1 926 544	37.2559
1164	1 354 896	34.1174	1239	1 535 121	35.1994	1314	1 726 596	36.2491	1389	1 929 321	37.2693
1165	1 357 225	34.1321	1240	1 537 600	35.2136	1315	1 729 225	36.2629	1390	1 932 100	37.2827
1166	1 359 556	34.1467	1241	1 540 081	35.2278	1316	1 731 856	36.2767	1391	1 934 881	37.2961
1167	1 361 889	34.1614	1242	1 542 564	35.2420	1317	1 734 489	36.2905	1392	1 937 664	37.3095
1168	1 364 224	34.1760	1243	1 545 049	35.2562	1318	1 737 124	36.3043	1393	1 940 449	37.3229
1169	1 366 561	34.1906	1244	1 547 536	35.2704	1319	1 739 761	36.3180	1394	1 943 236	37.3363
1170	1 368 900	34.2053	1245	1 550 025	35.2846	1320	1 742 400	36.3318	1395	1 946 025	37.3497
1171	1 371 241	34.2199	1246	1 552 516	35.2987	1321	1 745 041	36.3456	1396	1 948 816	37.3631
1172	1 373 584	34.2345	1247	1 555 009	35.3129	1322	1 747 684	36.3593	1397	1 951 609	37.3765
1173	1 375 929	34.2491	1248	1 557 504	35.3270	1323	1 750 329	36.3731	1398	1 954 404	37.3898
1174	1 378 276	34.2637	1249	1 560 001	35.3412	1324	1 752 976	36.3868	1399	1 957 201	37.4032
1175	1 380 625	34.2783	1250	1 562 500	35.3553	1325	1 755 625	36.4005	1400	1 960 000	37.4166

APPENDIX II

LIST OF SOME RELATED PUBLICATIONS ON QUALITY CONTROL

Texts

Bennett, C. A. and Franklin, N. L., *Statistical Analysis in Chemistry and the Chemical Industry*, Wiley, New York, 1954.
Bowker, A. H. and Lieberman, G. L., *Engineering Statistics*, 2nd ed., Prentice-Hall, Englewood Cliffs, N. J., 1972.*
Burr, I. W., *Statistical Quality Control Methods*, Marcel Dekker, Inc., New York, 1976.*
Cramer, H., *Mathematical Methods of Statistics*, Princeton University Press, Princeton, N. J., 1946.
Dixon, W. J. and Massey, F. J., Jr., *Introduction to Statistical Analysis*, 3rd ed., McGraw-Hill, New York, 1969.
Duncan, A. J., *Quality Control and Industrial Statistics*, 4th ed., Richard D. Irwin, Inc., Homewood, Ill., 1974.*
Feller, William, *An Introduction to Probability Theory and Its Application*, 3rd ed., Wiley, New York, 1968.
Grant, E. L. and Leavenworth, R. S., *Statistical Quality Control*, 4th ed., McGraw-Hill, New York, 1972.*
Hald, A., *Statistical Theory with Engineering Applications*, Wiley, New York, 1952.
Hoel, P. G., *Introduction to Mathematical Statistics*, 4th ed., Wiley, New York, 1971.
Juran, J. M., *Quality-Control Handbook*, 3rd ed., McGraw-Hill, New York, 1974.*
Mood, A. M. and Graybill, F. A., *Introduction to the Theory of Statistics*, 2nd ed., McGraw-Hill, New York, 1963.
Moroney, M. J., *Facts from Figures*, 3rd ed., Penguin, Baltimore, Md., 1956.
Ott, Ellis, *Process, Quality Control*, McGraw-Hill, New York, 1975.*
Rickmers, A. D. and Todd, H. N., *Statistics—An Introduction*, McGraw-Hill, New York, 1967.*
Shewhart, W. A., *Economic Control of Quality of Manufactured Product*, Van Nostrand, New York, 1931.
Shewhart, W. A., *Statistical Method from the Viewpoint of Quality Control*, Graduate School of the U. S. Department of Agriculture, Washington, D. C., 1939.*
Simon, L. E., *An Engineer's Manual of Statistical Methods*, Wiley, New York, 1941.*
Tippett, L. H. C., *Technological Applications of Statistics*, Wiley, New York, 1950.

Journals

Annals of Probability
Annals of Statistics
Applied Statistics
Biometrika
Journal of the American Statistical Association
*Journal of Quality Technology**
Journal of the Royal Statistical Society, Series B

*With special reference to quality control.

*Quality Progress**
Reports of Statistical Application Research
*Technometrics**

Pamphlets

ANSI Standard Z1.1-1958 (R 1975), "Guide for Quality Control" (ASQC Standard B1)
ANSI Standard Z1.2-1958 (R 1975), "Control Chart Method for Analyzing Data" (ASQC Standard B2)
ANSI Standard Z1.3-1958 (R 1975), "Control Chart Method of Controlling Quality During Production" (ASQC Standard B3)
ASQC General Publications

APPENDIX III

Designation: E 122 - 72

American National Standard Z 1.11-1973
Approved July 25, 1973
By American National Standards Institute

Standard Recommended Practice for
CHOICE OF SAMPLE SIZE TO ESTIMATE THE AVERAGE QUALITY OF A LOT OR PROCESS[1]

This Standard is issued under the fixed designation E 122; the number immediately following the designation indicates the year of original adoption or, in the case of revision, the year of last revision. A number in parentheses indicates the year of last reapproval.

1. Scope

1.1 This recommended practice presents simple methods for calculating how many units to include in a sample in order to estimate, with a prescribed precision, the average of some characteristic for all the units of a lot of material, or the average produced by a process.

2. Symbols

E = maximum allowable error between the estimate to be made from the sample and the result of measuring (by the same methods) all the units in the lot or process.

e = $E/\bar{X}'$

$\bar{X}'$ = lot or process mean or expected value of X, the result of measuring all the units in the lot or process.

$\bar{X}'_0$ = an advance estimate of $\bar{X}'$.

N = size of the lot.

n = size of the sample taken from infinite lot.

n_L = size of the sample from a finite lot.

p' = true (unknown) fraction of a lot or process whose units have the characteristic under investigation.

p'_0 = an advance estimate of p'.

p = fraction defective in the sample.

R = range of a set of sampling values. The largest minus the smallest observation.

$\bar{R}$ = average of the ranges of several samples, all of the same size.

σ' = (unknown) standard deviation of the units of a lot or process.

σ'_0 = an advance estimate of σ'.

s = $[\Sigma(X_i - \bar{X})^2]/[(n - 1)]^{1/2}$, an estimate of the standard deviation σ' defined by the formula (with appropriate degrees of freedom).

$\bar{s}$ = average s from several samples.

V' = $\sigma'/\bar{X}'$, the coefficient of variation of the lot or process.

V'_0 = an advanced estimate of V'.

V = $s/\bar{X}$, or $\bar{R}/d_2\bar{X}$, the coefficient of variation in the sample.

X = random variable or numerical value of the characteristic of an individual unit being measured.

$\bar{X}$ = average of X from several observations = $\Sigma x_i/n$.

3. Empirical Knowledge Needed

3.1 Some empirical knowledge of the problem is desirable in advance, as follows:

3.1.1 We may have some idea about the standard deviation of the characteristic.

3.1.2 If we have not had enough experience to give a narrow band estimate for the standard deviation, we may be able to state our belief about the range or spread of the characteristic from its lowest to its highest value and possibly about the shape of the distribution of the characteristic; for instance, we might be able to say whether most of the values lie at one end of the range, or are mostly in the middle, or run rather uniformly from one end to the other (see Section 7).

3.2 If the aim is to estimate the fraction

[1] This recommended practice is under the jurisdiction of ASTM Committee E-11 on Statistical Methods.

Current edition approved April 3, 1972. Published June 1972. Originally published as E 122 – 56 T. Last previous edition E 122 – 58.

defective, then each unit has a value of 0 or 1 (not defective or defective), and the standard deviation as well as the shape of the distribution depends only on p', the fraction defective of the lot or process. Some rough idea concerning the size of p' is therefore needed, which may be derived from preliminary sampling or from previous experience.

3.3 Sketchy knowledge is sufficient to start on, although more knowledge permits greater economy in the sample. Seldom will there be difficulty in acquiring enough information to compute the required size of sample. A sample that is larger than the equations indicate is used in actual practice when the empirical knowledge is only sketchy to start with, and if the desired precision is critical. Extra insurance is the price of incomplete knowledge.

3.4 In any case, even when starting with sketchy knowledge, the precision of the estimate made from a random sample may itself be estimated from the sample. This estimation of the precision reached by the first sample makes it possible to fix more economically the sample size for the next sample of a similar material. In other words, information concerning the process, and the material produced thereby, accumulates and should be used.

4. Precision Desired

4.1 The approximate precision desired for the estimate must be prescribed. That is, it must be decided what maximum deviation E, can be tolerated between the estimate to be made from the sample and the result that would be obtained by measuring every unit in the lot or process.

5. Equations for Calculating Sample Size

5.1 The equation for the size, n, of the sample is as follows:

$$n = (3\sigma'_0/E)^2 \qquad (1)$$

where:

σ'_0 = the advance estimate of the standard deviation of the lot or process,

E = the maximum allowable error between the estimate to be made from the sample and the result of measuring (by the same methods) all the units in the lot or process, and

3 = a factor corresponding to a very low probability that the difference between the sample estimate and the result of measuring (by the same methods) all the units in the lot or process is greater than E. The choice of the factor 3 is recommended for general use. With the factor 3, and with a lot or process standard deviation equal to the advance estimate, it is "practically certain" that the sampling error will not exceed E. Where a lesser degree of certainty is desired, a smaller factor may be used.

NOTE 1—For example, the factor 2 in place of 3 gives a probability of about 45 parts in 1000 that the sampling error will exceed E. Although distributions met in practice may not be normal, the following table (based on the normal distribution) indicates approximate probabilities:

Factor	Approximate Probability
3	0.003 or 3 in 1000
2.58	0.010 or 10 in 1000
2	0.045 or 45 in 1000
1.96	0.050 or 5 in 100 (1 in 20)
1.64	0.100 or 10 in 100 (1 in 10)

NOTE 2—If a lot of material has a highly skewed distribution in the characteristic measured, the factor 3 will give a different probability, possibly as great as 16 parts in 1000. If there is anxiety about the effect of skewness, there are two things to do:

(*1*) Probe the material with a view to discovering, for example, extra-high values, or possibly spotty runs of abnormal character, in order to approximate roughly the amount of the skewness, for use with statistical theory and adjustment of the sample size if necessary.

(*2*) Search the lot for abnormal material and segregate it for separate treatment.

5.2 It is sometimes convenient to use Eq 1 in another form as follows:

$$n = (3V'_0/e)^2 \qquad (2)$$

where:

V'_0 (coefficient of variation) = $\sigma'_0/\bar{X}'_0$, the advance estimate of the coefficient of variation, expressed in percent (or as a fraction),

e = $E/\bar{X}'$, the allowable sampling error expressed as a percent (or fraction) of $\bar{X}'$, and

$\bar{X}'$ = the expected value of the characteristic being measured.

5.2.1 There are some materials for which σ' varies approximately with $\bar{X}'$, in which case $V'(= \sigma'/\bar{X}')$ remains approximately constant from large to small values of $\bar{X}'$. If the relative error, e, is to be the same for all values of $\bar{X}'$, then everything on the right-hand side of Eq 2 is a constant; hence n is also a constant, which means that the same sample size n would be required for all values of $\bar{X}'$.

5.3 If the problem is to estimate the lot fraction defective, then $(\sigma'_0)^2$ is replaced by $p'_0(1 - p'_0)$, so that Eq 1 becomes:

$$n = (3/E)^2 p'_0 (1 - p'_0) \quad (3)$$

where:

p'_0 = the advance estimate of the lot or process fraction defective p'.

If p'_0 is small, so that $p'_0 n$ is less than 4, then 3.25 should be used in place of 3 in Eq 3 to compensate for the skewness of the p distribution for small values of p'.

5.4 When the average of a particular lot of limited size is needed, and an estimate of the average for the process is not part of the problem, the required sample size is less than Eqs 1, 2, and 3 indicate. The sample size for estimating the average of the finite lot will be:

$$n_L = [N/(N + n)]n \quad (4)$$

where:

n = the value computed from Eqs 1, 2, or 3, and

N = the lot size.

This reduction in size is usually of little importance unless n is 10 percent or more of N.

6. Reduction of Empirical Knowledge to a Numerical Value of σ'_0 (Data for Previous Samples Available)

6.1 This section illustrates the use of the equations in Section 4 when there are data for previous samples.

6.2 *For Eq 1*—Compute the standard deviation $s = \sqrt{\Sigma(X_i - \bar{X})^2/(n - 1)}$ for several samples, and use the average of them (Notes 3 and 4) if they are not too dissimilar, for σ'_0 (Note 4).

NOTE 3—A simple way to derive a value for σ'_0 from sample data is to arrange the observed values in a random order, and then average the ranges of successive groups of 4, 5, 8, or 10 observed values. (Normal theory shows that the optimum subgroup size is 8.) If $\bar{R}$ is the average of these ranges, then $\bar{R}/d_2$ is used for σ'_0.

The following table shows some selected values of d_2:[2]

Group Size	d_2
2	1.13
4	2.06
5	2.33
8	2.85
10	3.08

NOTE 4—For small samples a good average of k samples of s's is obtained by taking $\bar{s} = \Sigma c_4 s/k$ where the c_4 in each case is given by Table 1 of Definitions, Symbols, Formulas, and Tables for Quality Control (ASQC Standard A1).[3]

6.2.1 **Example 1**—*Use of* $\bar{s}$:

6.2.1.1 *Problem*—To compute the sample size needed to estimate the average transverse strength of a lot of bricks when the desired value of E is 50 psi.

6.2.1.2 *Solution*—From the data of three previous lots, the values of estimated standard deviation were found to be 215, 192, and 202 psi, based on samples of 100 bricks. The average of these three standard deviations is 203 psi, whence Eq 1 gives the following equation:

$$n = [(3 \times 203)/50]^2 = (12.2)^2 = 149 \text{ bricks}$$

for the required size of sample to give a maximum sampling error of 50 psi.

6.3 *For Eq 2*—If σ' varies approximately with $\bar{X}'$ for the characteristic of the material to be measured, compute both the average, $\bar{X}$, and the standard deviation, s, for several samples (unless they are already available). An average (Note 5) of the several values of $V = s/\bar{X}$, if they are not too dissimilar, may be used for V'_0.

NOTE 5—A good procedure to use when the sample sizes are small is to take $1/V'_0 = [(n_1/V_1C_1) + (n_2/V_2C_2) + \cdots (n_k/V_kC_k)]/(n_1 + n_2 + \cdots n_k)$ where values for the C_k are as follows:

n	C
3	1.722
4	1.383
5	1.253
6	1.189
7	1.151
8	1.126
9	1.108
10	1.094
17	1.050
33	1.024
∞	1.000

These values of C are based on the fact that $\sqrt{n}/V$ has a noncentral t-distribution.

6.3.1 **Example 2**—*Use of* V, *the estimated coefficient of variation:*

6.3.1.1 *Problem*—To compute the sample size needed to estimate the average abrasion resistance of a material when the desired value of e is 0.10 or 10 percent.

6.3.1.2 *Solution*—There are no data from

---

[2] See *Manual on Quality Control of Materials, ASTM STP 15-C*, ASTTA, Am. Soc. Testing Mats., 1951, p. 63.

[3] Available from American Society for Quality Control, 161 W. Wisconsin Ave., Milwaukee, Wis. 53203.

previous samples of this same material, but data for six samples of similar materials show a wide range of resistance. However, the values of estimated standard deviation are approximately proportional to the observed averages, as shown in the following table:

Lot No.	Sample Sizes	Average Cycles	Observed range, R	Estimate of σ' = $R/3.08$[a]	Coefficient of Variation, percent
1	10	90	40	13.0	14
2	10	190	100	32.5	17
3	10	350	140	45.5	13
4	10	450	220	71.4	16
5	10	1000	360	116.9	12
6	10	3550	2090	678.6	19
Avg					15.2

[a] Values of standard deviation, s, may be used instead of the estimates made from the range, if they are preferred or already available. The use of the average of the observed values of the coefficient of variation for V'_0 in Eq 2 gives the following:

$$n = [(3 \times 15.2)/10]^2 = (4.6)^2 = 21.2\text{—}22 \text{ specimens}$$

for the required size of sample to give a maximum sampling error of 10 percent of the expected value.

6.3.1.3 If a maximum allowable error of 5 percent were needed, the required sample size would be 85 specimens. The data supplied by the prescribed sample will be useful for the study in hand, and also for the next investigation of similar material.

6.4 *For Eq 3*—Compute the estimated fraction defective, p, for each sample. Then for the weighted average use the following equation:

$$\bar{p} = \frac{\text{total number of defectives in all samples}}{\text{total number of units in all samples}}$$

6.4.1 **Example 3**—*Use of p:*

6.4.1.1 *Problem*—To compute the size of sample needed to estimate the fraction defective in a lot of alloy steel track bolts and nuts when the desired value of E is 0.04.

6.4.1.2 *Solution*—The data in the following table from four previous lots were used for an advance estimate of p':

Lot No.	Sample Size	No. of Defectives	Fraction Defective
1	75	3	0.040
2	100	10	0.100
3	90	4	0.044
4	125	4	0.032
Total	390	21	

$$\bar{p} = 21/390 = 0.054$$

$$n = (3/0.04)^2(0.054)(0.946) = [(9 \times 0.0511)/0.0016] = 287.4 = 288$$

If the desired value of E were 0.01, the required sample size would be 4600. The required sample would be smaller if Eq 4 applies.

7. Reduction of Empirical Knowledge to a Numerical Value for σ'_0 (No Data from Previous Samples of the Same or Like Material Available)

7.1 This section illustrates the use of the equations in Section 5 when there are no actual observed values for the computation of σ_0.

7.2 *For Eq 1*—From past experience, try to discover what the smallest and largest values of the characteristic are likely to be. (If this is not known, obtain this information from some other source.) Try to picture how the other observed values may be distributed. A few simple observations and questions concerning the past behavior of the process, the usual procedure of blending, mixing, stacking, storing, etc., and knowledge concerning the aging of material and the usual practice of withdrawing the material (last in, first out; or last in, last out) will usually elicit sufficient information to distinguish between one form of distribution and another in Fig. 1 (Note 6). In case of doubt, or in case the desired precision E is a critical matter, the rectangular distribution may be used. The price of the extra protection afforded by the rectangular distribution is a larger sample, owing to the larger standard deviation thereof. At the worst, if the isosceles triangle is used when the other triangular or the rectangular distributions are better descriptions, then the standard deviation obtained is larger by no more than 40 percent, as shown by comparing the formulas for the standard deviations given in Fig. 1. The sizes of subsequent samples may then be adjusted upward, if necessary.

NOTE 6—The standard deviation of the normal distribution in Fig. 1 is a safe assumption for materials with a good history of control, in which case an advance estimate of σ' would be available.

7.2.1 The standard deviation estimated from one of the formulas of Fig. 1 may be used as an advance estimate of σ' in Eq 1. This method of advance estimation is in constant use and is often preferable to doubtful observed values.

7.2.2 **Example 4**—*Use of σ'_0 from Fig. 1:*

7.2.2.1 *Problem* (*Same as Example 1*)—To compute the sample size needed to estimate the average transverse strength of a lot of bricks when the desired value of E is 50 psi.

7.2.2.2 *Solution*—From past experience the spread of values of transverse strength for a lot of bricks has been about 1200 psi. The values were heaped up in the middle of this band, but not necessarily normally distributed.

7.2.2.3 The isosceles triangle distribution in Fig. 1 appears to be most appropriate; the advance estimate of σ' is 1200/4.9 = 245 psi. Then:

$$n = [(3 \times 245)/50]^2$$
$$= (14.7)^2 = 216.1 = 217 \text{ bricks}$$

7.2.2.4 The difference in sample size between 217 and 149 bricks (found in Example 1) is the price of sketchy knowledge.

7.3 *For Eq 2*—While the estimation of the coefficient of variation of a universe by use of Fig. 1 is possible, it is not recommended. In general, the knowledge that the use of V'_0 instead of σ'_0 is preferable would be obtained from the analysis of actual data, in which case the methods of Section 6 apply.

7.4 *For Eq 3*—From past experience, estimate approximately the band within which the fraction defective is likely to lie. Turn to Fig. 2 and read off the value of $(\sigma')^2 = p'(1 - p')$ for the middle of the possible range of p', and use it in Eq 3. In case the desired precision is a critical matter, use the largest value of $(\sigma')^2$ within the possible range of p'.

8. Consideration of Cost

8.1 After the required size of sample to meet a prescribed precision is computed from Eqs 1, 2, or 3, the next step is to compute the cost of testing this size of sample. If the cost is too great, it may be possible to relax the required precision (or the equivalent, which is to accept an increase in the probability (Section 5) that the sampling error may exceed the maximum allowable error, E) and to reduce the size of the sample to meet the allowable cost.

8.2 Equation 1 gives n in terms of a prescribed precision, but we may solve it for E in terms of a given n and thus discover what precision is possible for a given allowable cost. The same may be done for Eqs 2 and 3.

8.3 It is necessary to specify either the desired allowable error, E, or the allowable cost; otherwise there is no proper size of sample.

9. Selection of the Sample

9.1 In order to make any estimate for a lot or for a process, on the basis of a sample, it is necessary to select the units in the sample at random. Randomness is not just accident or lack of direction; it is the product of a definite and willful effort to produce disorder. The only universally acceptable definition of a random selection is one made by the use of random numbers, which in effect is the guarantee of thorough stirring of the sampling units in a lot.

9.2 In the use of random numbers, the material must first be broken up in some manner into "sampling units." Moreover, each sampling unit must be identifiable by a serial number, actual or by some rule. For packaged articles, a rule is easy; the package contains a certain number of articles in definite layers, arranged in a particular way, and it is easy to devise some system for numbering the articles. In the case of bulk material like ore, or coal, or a barrel of bolts or nuts, the problem of defining usable sampling units must take place at an earlier stage of manufacture.

9.3 It is not the purpose here to discuss the handling of materials, nor to find ways by which one can with surety discover the way to a satisfactory type of sampling unit. Instead, it is assumed that a suitable sampling unit has been defined, and then the aim is to answer the question of how many to draw.

ASTM E 122

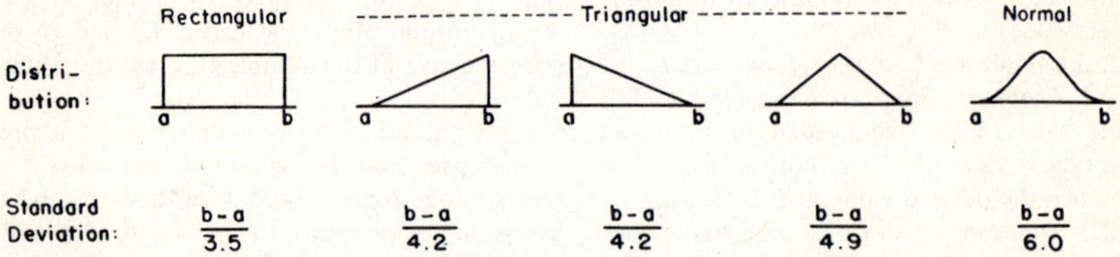

FIG. 1 Some Types of Distributions and Their Standard Deviations.

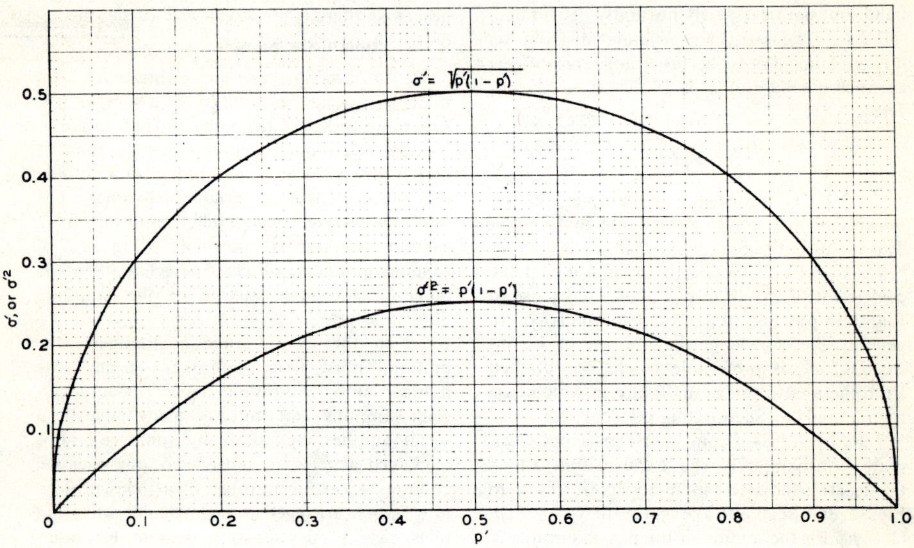

FIG. 2 Values of σ', or (σ')², Corresponding to Values of p'.

By publication of this standard no position is taken with respect to the validity of any patent rights in connection therewith, and the American Society for Testing and Materials does not undertake to insure anyone utilizing the standard against liability for infringement of any Letters Patent nor assume any such liability.